An Introductory
Global CO$_2$ Model

An Introductory

Global CO$_2$ Model

with Companion Media Pack

Anthony J McHugh

Lehigh University, USA

Graham W Griffiths

City University London, UK

William E Schiesser

Lehigh University, USA

World Scientific

NEW JERSEY · LONDON · SINGAPORE · BEIJING · SHANGHAI · HONG KONG · TAIPEI · CHENNAI

Published by

World Scientific Publishing Co. Pte. Ltd.

5 Toh Tuck Link, Singapore 596224

USA office: 27 Warren Street, Suite 401-402, Hackensack, NJ 07601

UK office: 57 Shelton Street, Covent Garden, London WC2H 9HE

Library of Congress Cataloging-in-Publication Data
McHugh, Anthony J.
  An introductory global CO2 model / Anthony J. McHugh (Lehigh University, USA), Graham W. Griffiths (City University London, UK) & William E. Schiesser (Lehigh University, USA).
    pages cm
  ISBN 978-9814663038
  1. Atmospheric carbon dioxide--Mathematical models. 2. Atmospheric carbon dioxide--Climatic factors. 3. Atmospheric carbon dioxide--Environmental aspects. 4. Carbon dioxide--Environmental aspects. I. Griffiths, Graham W. II. Schiesser, W. E. III. Title. IV. Title: Introductory global carbon dioxide model.
  QC879.8.M44 2015
  551.51'12--dc23

                                                                    2015009064

British Library Cataloguing-in-Publication Data
A catalogue record for this book is available from the British Library.

In-house Editors: Prathima/Amanda Yun

Typeset by Stallion Press
Email: enquires@stallionpress.com

Printed in Singapore

To Svante Arrhenius and Charles Keeling

CONTENTS

PREFACE

The consequences of rising concentrations of atmospheric CO_2 are receiving increased attention in the news media as a major contributor to extreme climate change and global warming. For example, from the *New York Times* [12] and restated in Appendix D of this book:

> "Scientific monitors reported that the gas (CO_2) had reached an average daily level that surpassed 400 parts per million (ppm) — just an odometer moment in one sense, but also a sobering reminder that decades of efforts to bring human-produced emissions under control are faltering. The best available evidence suggests the amount of the gas in the air has not been this high for at least three million years, before humans evolved, and scientists believe the rise portends large changes in the climate and the level of the sea."

And from [14], "Climate-related disasters around the world (include): hurricanes and tornadoes, droughts and wildfires, extreme heat waves and equally extreme cold, rising sea levels and floods. Even when people have doubts about the causal relationship of global warming to these episodes, they cannot help being psychologically affected. Of great importance is the growing recognition that the danger encompasses the entire earth and its inhabitants. We are all vulnerable."

However, the CO_2 problem is not well understood quantitatively by a general audience. Although some measures of atmospheric CO_2 accumulation have been reported such as the 400 ppm from the 10 May 2013 *New York Times* noted above, other measures are less well understood, e.g., the increasing rate of CO_2 emissions, the movement of carbon to and from other parts of Earth's environmental system, particularly the oceans with accompanying acidification.

A mathematical model can be particularly informative and helpful for understanding what is happening. We therefore present an introductory global CO_2 model that gives some key numbers, for example, atmospheric CO_2 concentration in ppm and ocean pH as a function of time for the calendar years 1850 (preindustrial) to 2100 (a modest projection into the future).

The model is based on just seven ordinary differential equations (ODEs) and is therefore intended as an introduction to some basic concepts and as a starting point for more detailed studies. The ODEs are carbon balances for seven reservoirs: upper atmosphere, lower atmosphere, long-lived biota, short-lived biota, ocean upper layer, ocean deep layer and marine biosphere.

Basic calculus is the only required mathematical background, e.g., derivatives and integration, so that the model is accessible to high school students as well as beginning college and university students. Specifically, the ODEs define derivatives of the form dy/dt where y is a dimensionless reservoir carbon concentration and t is time as a calendar year over the interval $1850 \leq t \leq 2100$. The ODEs are integrated numerically to give $y(t)$. The solutions are presented numerically and graphically. The essential mathematical concepts are embedded in the fundamental theorem of calculus:

$$\int_{t_0}^{t} (dy/dt) \, dt = y(t) - y(t_0),$$

where $t_0 = 1850$ and $y(t_0)$ is the ODE initial condition (IC). The integration is carried out numerically by library initial-value ODE integrators.

Some background in computer programming would also be helpful, but not essential. The programming of the model is in R^1 and Matlab,[2] two basic, widely used scientific programming systems. Thus, it can be executed on modest computers and is generally

---

[1]R is an open source scientific programming system that can be downloaded (no cost) from the Internet.

[2]Matlab is a commercial product available from the Mathworks, Natick, MA. Octave is an open source (no cost) alternative that will run the Matlab code with minimal changes.

accessible and usable worldwide. The book could also be used by readers interested in Matlab and/or R programming or a translation of one to the other. This additional use is facilitated by a side-by-side format of sections of R and Matlab code with a comparison of the two.

The model includes ocean chemistry calculations that address acidification with ocean pH as an output, typically ranging from 8.2 to 7.8 (pH decreases with increasing acidity). These calculations illustrate some basic numerical procedures, such as a Newton solver applied to a fourth-order polynomial to calculate ocean pH and spline interpolation to provide additional model outputs. The problem of acidification has important implications for a major food source, deterioration of coral and the associated marine life, as well as ocean CO_2 uptake [9].

A basic global warming component has been added based on CO_2 buildup in the lower atmosphere but since climate change (e.g., warming) is still a controversial and uncertain area, the primary focus is on carbon buildup in the atmosphere and oceans which is being measured quantitatively and is therefore undisputed.

Our intention is to provide a quantitative introduction to the CO_2 problem so that the model is basic and educational, and is not a state-of-the-art research model. We hope that the model serves this educational purpose, and we welcome comments, questions about model implementation and use, and suggestions for improvements (directed to wes1@lehigh.edu).

Anthony J. McHugh
Graham W. Griffiths
William E. Schiesser
London, UK
and
Bethlehem, PA, USA
July 1, 2015

1 INTRODUCTION

The anticipated changes in the Earth's climate that are now widely discussed are due in large part to the accumulation of so-called greenhouse gases (GHGs) in the Earth's atmosphere. The increase in GHGs causes a reduction in the re-radiation of energy from the Sun back into outer space. Since less energy leaves the Earth's atmosphere, heating of the atmosphere results in a temperature rise. This temperature rise, so-called global warming, is in turn a driving force for climate change.

Carbon dioxide (CO_2) is the major GHG, with increasing levels arising primarily from the burning of fossil fuels. Thus, changes in CO_2 level or concentration in the Earth's atmosphere is of paramount importance in understanding anticipated warming and climate change. A second aspect of CO_2 accumulation in the atmosphere that is not as generally recognized and appreciated as temperature rise is the accumulation of carbon (from CO_2) in the oceans that leads to ocean acidifcation. CO_2 dissolves in ocean water and undergoes a series of chemical changes that ultimately leads to increased hydrogen ion concentration, denoted subsequently as $[H^+]$, and thus acidification. This increase in $[H^+]$ is manifested as a decrease in pH; note that $[H^+]$ and pH move in opposite directions due to the basic relation[1]

$$pH = -log_{10}[H^+]. \tag{1.1}$$

---

[1] log_{10} indicates a base-10 logarithm. To briefly review, a number (real, nonnegative) can be expressed as $c = b^a$ where a is the logarithm (log) of c to the base b (e.g., 10; $e = 2.718282$, the base of the "natural" logarithm system). An important application of logs is to facilitate the multiplication or division

For example, if $[H^+] = 10^{-8}$, $pH = 8$ while if $[H^+] = 10^{-7}$, $pH = 7$.

As a point of notation, square brackets placed around a chemical formula, e.g., $[H^+]$, denotes a molar concentration. The units of concentration are typically mols/liter, millimols/liter or micromols/liter (where liter is a liter of aqueous solution). Since for H_2O one liter weighs one kilogram (kg) (because the density of H_2O is $1\,g/$milliliter $= 1\,g/cc$ with 1 liter $= 1000$ milliliters), these concentrations are also in reciprocal kg. For the purpose of using Eq. (1.1) to calculate pH, $[H^+]$ is in $g\,$mol/liter $= kg\,$mol/m^3.

Also, mols are taken specifically as g mols. One g mol of a chemical quantity has a weight equal to its atomic or molecular weight in g. For example, H_2O has a molecular weight of $2(1) + 16 = 18$ g. Thus, one g mol of H_2O is $18\,$g.

The causes for changing environmental CO_2 levels are complex and not completely understood. But increasing atmospheric CO_2 is clearly established through measurements over more than 50 years [8]. Enough is known about CO_2 accumulation[2] to begin to formulate quantitative descriptions of the various physical and chemical processes that determine CO_2 levels with the goal of projecting[3]

---

of numbers. For example, two numbers $c_1 = b^{a_1}, c_2 = b^{a_2}$ can be multiplied as $c_1 c_2 = b^{a_1} b^{a_2} = b^{a_1 + a_2}$, that is, the logs are merely added, followed by an antilog or inverse log (the common base b is raised to the power $a_1 + a_2$) to obtain the product. Similarly, for division, $c_1 / c_2 = b^{a_1} / b^{a_2} = b^{a_1} b^{-a_2} = b^{a_1 - a_2}$ so that the logs are subtracted to give the quotient c_1 / c_2.

[2]The literature pertaining to CO_2 is extensive and expanding rapidly so that any attempt at a comprehensive survey of the literature would necessarily be limited and soon out of date. Therefore, only references that are used in the subsequent discussion of the model are included in the bibliography which also includes sources that the reader can access for the most current information, for example, the Carbon Dioxide Information Analysis Center, http://cdiac.esd.ornl.gov/. Other general sources of information include [1, 5, 6, 7, 10, 11, 13, 16–20, 22, 28–30]. The emphasis of the following discussion is therefore on the description and computer implementation of the model.

[3]The word *projection* is used rather than *prediction* to suggest that the model output reflects a plausible future scenario and not a certain outcome in the future. Various scenarios can then be considered based on postulated changes in the model, e.g., the CO_2 emissions rate.

how atmospheric CO_2 levels might be expected to increase in the future. To this end, we describe here an introductory global CO_2 model that elucidates at a basic level the mechanisms which determine CO_2 buildup in the Earth's atmosphere.

The model is necessarily a simplification of the physical and chemical processes at work that determine CO_2 levels. However, the model provides insight into CO_2 dynamics (the variation of CO_2 levels over time); specifically, it can be used to study the effect of various phenomena and parameters that determine CO_2 levels, and how they change with time.

In particular, the effect of variations in the rate of anthropogenic emissions can be assessed. This is accomplished by the numerical integration of a system of ordinary differential equations (ODEs) starting with known conditions in the past (for example, at 1850, but this starting year can easily be changed). The forward integration of the ODE model equations through time can be to an arbitrary point in the future (for example, to 2100, but this final year can also easily be changed). The details of the ODE model, and some representative output from the model, are discussed subsequently.

A major advantage of a computer-based mathematical model is the execution of the associated computer code[4] for the solution of the model equations to observe the effect of postulated conditions, e.g., the rate of anthropogenic CO_2 emissions. Thus, although only a limited set of model outputs is considered here, the code can easily be executed for other conditions to observe the effect of the model structure and parameters on projected CO_2 levels. Ideally, this process should elucidate the most relevant and sensitive conditions that determine future CO_2 levels, and thereby give an indication of plausible future CO_2 changes.

The model focuses only on CO_2. It does not have a climate component and it has only a basic global warming component consisting

---

[4]The coding is for two well established scientific computing systems: (1) Matlab, a commercially available programming system and (2) R, an open-source programming system available from the Internet. The details for using Matlab and R as applied to the CO_2 model are subsequently discussed in detail.

of multiplication of atmospheric CO_2 (in ppm) by a temperature sensitivity. While the current levels of CO_2 are relatively well known from measurements, the net effect of CO_2 on the Earth's climate is not well understood quantitatively at this point in time. Thus we have limited the model to CO_2 dynamics as a main determinant of global warming and climate, but we do not attempt to explain the resulting dynamics of anticipated global warming and climate change.

The model does, however, have as an output ocean H^+ and pH as a function of time. As with atmospheric CO_2, the long-term effects of ocean acidification are not completely understood at this point in time. However, two relatively well established effects can be observed:

- The ocean pH is decreasing and this can be measured.
- The effect of acidification on coral is being observed. We take $CaCO_3$ as the main component of coral, and the basic chemistry relating $CaCO_3$ and H^+ is considered toward the end of this discussion.

Of course, other important effects of acidification could be considered, for example, on the ocean biosphere, but they are not included in the model. No doubt these effects will be important and will be elucidated with future research.

2 MODEL STRUCTURE

The model structure [2], [27] is illustrated in Fig. 1.

- Seven reservoirs are used to represent the global CO_2 system:

  1. Upper atmosphere
  2. Lower atmosphere
  3. Short-lived biota
  4. Long-lived biota
  5. Upper ocean layer
  6. Deep ocean layer
  7. Marine biosphere

  These seven reservoirs were selected because each is considered to have an important effect on the global CO_2 dynamics. However, reservoirs can easily be added (or removed) from the mathematical model that is discussed subsequently.

- Within each reservoir, we consider CO_2 (and related carbon concentrations). As the most basic approach, we consider only one uniform CO_2 concentration in each reservoir. In other words, we neglect any spatial variation in CO_2 concentration. This assumed uniformity in space is, of course, an idealization and major research efforts are underway to elucidate spatial variations in CO_2 concentrations, particularly in the oceans. However, for our purpose of developing a basic model, we will assume spatial uniformity, sometimes termed perfect mixing or complete mixing [27].

- Thus, the CO_2 concentration in each reservoir varies only with time, and we are therefore using ODEs in time as the basic mathematical structure of the model (generally, ODEs have only one independent variable which in our analysis is time, denoted

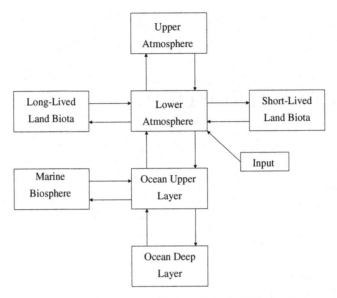

Figure 1: Diagram of the global CO_2 model

as t). Integration of the ODE system with respect to t gives the time variation of the reservoir concentrations, starting from an initial condition specified at the calendar year 1850.

- The time variation of the reservoir concentrations is driven through time by the anthropogenic emission of CO_2 to the lower atmosphere, identified as "input" in Fig. 1. This input is generally termed an emission rate.

3 MODEL EQUATIONS

As a significant simplification, only one CO_2 concentration in each reservoir is considered (by the perfect mixing assumption [27]), and the seven concentrations are identified mathematically as

Table 1: Model dependent variable concentrations

Concentration	Reservoir
$y_{ua}(t)$	upper atmosphere
$y_{la}(t)$	lower atmosphere
$y_{sb}(t)$	short-lived biota
$y_{lb}(t)$	long-lived biota
$y_{ul}(t)$	ocean upper layer
$y_{dl}(t)$	ocean deep layer
$y_{mb}(t)$	marine biosphere
t	time (calendar year)

The concentrations in Table 1 have the dimensionless form

$$y(t) = \frac{c_{\dim}(t) - c_{\dim}(t = 1850)}{c_{\dim}(t = 1850)}, \tag{3.1}$$

where $c_{\dim}(t)$ is a dimensional concentration with units, e.g., for the upper and lower atmospheres, the dimensional concentrations might be in parts per million, ppm (interpreted as μatm), or GtC (gigatons carbon $= 10^9$ tons carbon). Some additional units and conversion factors follow:

Unit	Related unit
1 (metric) ton	10^3 kg
	10^6 gm
1 gigaton (= 1 Gt)	10^9 tons
1 petagram (= 1 Pg)	10^{15} gm
	1 Gt

Additional conversions follow from the preceding units. For example, 1 Gt $= 10^9$ tons $\times 10^6$ (g/ton) $= 10^{15}$ g $= 1$ Pg; in other words, 1 Gt $= 1$ Pg as noted above.

One important detail should be considered when working with a carbon model. The carbon can be included in various compounds, for example, CO_2, carbonic acid (in the oceans), H_2CO_3, bicarbonate, HCO_3^- and carbonate, CO_3^{2-}. When the masses of these various compounds are computed and compared, this should be done on a consistent basis, which is the mass of the carbon. For example, the mass of carbon in CO_2 is $12/(12 + 2(16)) = 12/44$ of the mass of the CO_2 (atomic weight carbon $= 12$, molecular weight of $CO_2 = 44$). The computed results, and data taken from the literature, should be stated and compared in consistent units, usually the mass of carbon, although the mass of CO_2 is used frequently in discussions of the upper and lower atmosphere.

Thus, Eq. (3.1) indicates the concentrations will be fractional changes from the initial dimensional concentration $c_{\text{dim}}(t = 1850)$. In other words, by the definition of Eq. (3.1), all of the initial concentrations will be zero since (with the initial time $t = 1850$)

$$y(t = 1850) = \frac{c_{\text{dim}}(t = 1850) - c_{\text{dim}}(t = 1850)}{c_{\text{dim}}(t = 1850)} = 0. \qquad (3.2)$$

Through the use of the dimensionless concentrations defined by Eq. (3.1), all seven reservoir concentrations of Table 1 are brought together on a common scale (with the seven concentrations starting from their zero initial values at $t = 1850$). This does not mean, however, that the model output (the seven dependent variables of Table 1) cannot also be expressed as dimensional concentrations for the purpose of analyzing the model output. This switch between dimensionless and dimensional variables is illustrated in the subsequent discussion.

We now consider the ODEs that, when integrated numerically, give the seven concentrations of Table 1. For the upper atmosphere, the ODE is

$$\frac{dy_{ua}}{dt} = \frac{1}{\theta_{la}^{ua}}(y_{la} - y_{ua}), \tag{3.3a}$$

$$y_{ua}(1850) = 0 \tag{3.3b}$$

(a more detailed discussion of Eqs. (3.3a) and (3.3b) is given in Appendix B).

Equation (3.3a) defines the derivative $\dfrac{dy_{ua}}{dt}$ which is the ratio of a differential change in the upper atmospheric CO_2, dy_{ua}, over a differential change in t, dt. The numerical solution of Eq. (3.3a) subject to initial condition (IC) (3.3b) is the dependent variable y_{ua} as a function of the independent variable t, $y_{ua}(t)$, in numerical form. Note that the LHS and RHS units of Eq. (3.3a) are consistent, that is, $1/time$ or $1/yr$ (since y_{ua} is dimensionless).

θ_{la}^{ua} is a mean residence time or mixing time for CO_2 that enters the upper atmosphere from the lower atmosphere (refer to the arrow of Fig. 1 for the lower atmosphere to upper atmosphere transfer of CO_2). Once θ_{la}^{ua} is specified, $y_{ua}(t)$ is completely defined by Eq. (3.3a) (when the initial condition (IC) $y_{ua}(1850) = 0$ of Eq. (3.3b) is also included). In other words, y_{ua} comes from the integration of Eq. (3.3a) (assuming y_{la} is available from the ODE for the lower atmosphere as discussed next; that is, the model is a set of simultaneous ODEs).

In naming the residence times, we have used the following general notation

$$\theta_{forcing\ variable}^{response\ variable}.$$

So, for example, in Eq. (3.3a), y_{la} is the forcing variable and y_{ua} is the response variable, so the residence time in Eq. (3.3a) is θ_{la}^{ua}.

To reiterate, Eq. (3.3a) states that the rate of transfer of CO_2 to the upper atmosphere, as expressed through the LHS derivative $\dfrac{dy_{ua}}{dt}$, is proportional to the difference between the fractional upper atmosphere carbon concentration (y_{ua}) and the fractional lower atmosphere carbon concentration (y_{la}), that is, the difference $y_{la} - y_{ua}$; the proportionality constant is $\dfrac{1}{\theta_{la}^{ua}}$ where θ_{la}^{ua} is the mean residence time.

Equation (3.3a) is a highly simplified mathematical description of the CO_2 transfer between the upper atmosphere and the lower atmosphere. It does, however, have several important properties:

- For $y_{la} > y_{ua}$, the net transfer of CO_2 is from the lower atmosphere to the upper atmosphere. In other words, $\dfrac{dy_{ua}}{dt} > 0$ as expected (y_{ua} will increase with t).
- Conversely, for $y_{ua} > y_{la}$, the net transfer of CO_2 is from the upper atmosphere to the lower atmosphere. In other words, $\dfrac{dy_{ua}}{dt} < 0$ as expected.
- At steady state (no changes in y_{ua} and y_{la} with t), $\dfrac{dy_{ua}}{dt} = 0$ as expected (when $y_{ua} = y_{la}$ in Eq. (3.3a)). In other words, this zero derivative means y_{ua} does not change with t.
- The rate of transfer of CO_2 between the upper and lower atmospheres is characterized in terms of just θ_{la}^{ua}. This is an important simplification. In other words, the difficult task of specifying the CO_2 fluxes between the upper and lower atmospheres is reduced to specifying just this single parameter (θ_{la}^{ua}). It might be argued that this is an oversimplification. But for the purposes of this introductory model, we accept this simplification with the understanding that some variation in θ_{la}^{ua} in multiple runs of the computer code will reveal just how sensitive the model output is to the particular value of θ_{la}^{ua}.
- Additionally, the dimensionless form of Eq. (3.3a) can be converted to dimensional form that leads to its interpretation as a basic mass balance. The details of this analysis are given in Appendix B.

The ODE for the lower atmosphere is developed in the same way as Eq. (3.3a). However, the ODE will be more complicated than

Eq. (3.3a) because of the multiple input/output streams of CO_2 (refer to Fig. 1). Now a mean residence time is required for each stream entering the lower atmosphere.

$$\frac{dy_{la}}{dt} = \frac{1}{\theta_{ua}^{la}}(y_{ua} - y_{la}) + \frac{1}{\theta_{sb}^{la}}(y_{sb} - y_{la}) + \frac{1}{\theta_{lb}^{la}}(y_{lb} - y_{la})$$

$$+ \frac{1}{\theta_{ul}^{la}}(y_{ul} - y_{la}) + Q_c(t), \tag{3.4a}$$

$$y_{la}(1850) = 0. \tag{3.4b}$$

Equation (3.4a) is stated on multiple lines to elucidate its structure. Note also: (1) $Q_c(t)$ is the anthropogenic CO_2 input of Fig. 1 and (2) the IC $y_{la}(1850) = 0$ (from Eq. (3.4b)) completes the mathematical specification for the calculation of y_{la} (assuming y_{ua} comes from Eq. (3.3a), and y_{sb}, y_{lb}, y_{ul} come from associated ODEs).

The ODEs for the other five reservoirs follow analogously from Eqs. (3.3) and (3.4). For the short-lived biota (Fig. 1), the equation is

$$\frac{dy_{sb}}{dt} = \frac{1}{\theta_{la}^{sb}}(y_{la} - y_{sb}), \tag{3.5a}$$

$$y_{sb}(1850) = 0 \tag{3.5b}$$

and for the long-lived biota

$$\frac{dy_{lb}}{dt} = \frac{1}{\theta_{la}^{lb}}(y_{la} - y_{lb}), \tag{3.6a}$$

$$y_{lb}(1850) = 0. \tag{3.6b}$$

The ODE for the ocean upper layer has three RHS terms to reflect the exchange with the lower atmosphere, ocean deep layer and marine biota (Fig. 1)

$$\frac{dy_{ul}}{dt} = \frac{1}{\theta_{la}^{ul}}(y_{la} - y_{ul}) + \frac{1}{\theta_{dl}^{ul}}(y_{dl} - y_{ul}) + \frac{1}{\theta_{mb}^{ul}}(y_{mb} - y_{ul}), \tag{3.7a}$$

$$y_{ul}(1850) = 0. \tag{3.7b}$$

Finally, the ODEs for the ocean deep layer and the marine biota are

$$\frac{dy_{dl}}{dt} = \frac{1}{\theta_{ul}^{dl}}(y_{ul} - y_{dl}), \tag{3.8a}$$

$$y_{dl}(1850) = 0, \tag{3.8b}$$

$$\frac{dy_{mb}}{dt} = \frac{1}{\theta_{ul}^{mb}}(y_{ul} - y_{mb}), \tag{3.9a}$$

$$y_{mb}(1850) = 0. \tag{3.9b}$$

Equations (3.3) to (3.9) constitute the complete ODE model for the seven dependent variables of Table 1. The one remaining detail is the source term, $Q_c(t)$, of Eq. (3.4b). This term will be taken as an exponential of the form

$$Q_c(t) = c_1 e^{r_1 t}, \tag{3.10}$$

where the constants c_1 and r_1 are selected in accordance with observed atmospheric variations in CO_2 concentration (as discussed subsequently). $Q_c(t)$ from Eq. (3.10) drives the entire model (seven ODE dependent variables) away from zero ICs and therefore the form of this CO_2 source term is a central issue in using the model.

In summary, the model of Eqs. (3.3) to (3.10) has the form of a system of linear, constant coefficient ODEs and is therefore of modest complexity to facilitate an introductory quantitative discussion of the CO_2 problem. This model format is based on the seminal paper of Bacastow and Keeling [2] and has been used subsequently by other researchers, e.g., [27].

4 MODEL PARAMETERS

The model parameters are: (1) the residence times in the preceding ODEs and (2) the constants c_1 and r_1 in Eq. (3.10). The residence times are summarized in Table 2 (their use in the ODE R and Matlab routines is discussed subsequently).

The emissions rate constants, c_1 and r_1 in Eq. (3.10) are set in R and Matlab functions as explained subsequently.

Table 2: Summary of reservoir residence times

Residence time	Value (yr)
θ_{ua}^{la}	5
θ_{sb}^{la}	0.75
θ_{lb}^{la}	150
θ_{ul}^{la}	30
θ_{la}^{ua}	3
θ_{la}^{sb}	0.75
θ_{la}^{lb}	150
θ_{la}^{ul}	80
θ_{dl}^{ul}	200
θ_{mb}^{ul}	5
θ_{ul}^{dl}	1500
θ_{ul}^{mb}	10

5 COMPUTER ROUTINES

We now consider a series of routines in R and Matlab for the solution of the model discussed in Chapter 3. The format is to list an R routine followed by the analogous Matlab routine, then give a comparative discussion of both routines. We start with the main programs.

5.1. Main Programs

The main R program is listed next (with some code indented or placed on two lines so that it fits on a page).

Listing 1a: Main program in R

```
#
# Delete previous workspaces
  rm(list=ls(all=TRUE))
#
# Access ODE integrator
  library("deSolve");
#
# Access files
  setwd("c:/model_R");
  source("model_1.R");
  source("splines.R");
  source("CO2_rate.R");
  source("plot_output.R");
#
# Declare (preallocate) arrays
  nout=26;
    r1e=rep(0,nout);  ppm=rep(0,nout); eps=rep(0,nout);
     Tc=rep(0,nout); hion=rep(0,nout);  pH=rep(0,nout);
    co2=rep(0,nout); hco3=rep(0,nout); co3=rep(0,nout);
  boh4=rep(0,nout);
#
```

```
# Dimensional ODE dependent variables
  cla=rep(0,nout);cua=rep(0,nout);csb=rep(0,nout);
  clb=rep(0,nout);cul=rep(0,nout);cdl=rep(0,nout);
  cmb=rep(0,nout);
#
# Select case (see CO2_rate.R)
  ncase=1;
#
# Equilibrium constants (in millimols/lt units)
  k0=0.03347e-03;
  k1=9.747e-04;
  k2=8.501e-07;
  kb=1.881e-06;
  kw=6.46e-09;
#
# Boron
  B=0.409;
#
# Initial conditions
  n=7;
  y0=rep(0,n);
#
# Independent variable for ODE integration
  t0=1850;tf=2100;
  tout=seq(from=t0,to=tf,by=(tf-t0)/(nout-1));
  ncall=0;
#
# Initialized carbon reservoirs (GtC)
  c0la=0.85*597;
  c0ua=0.15*597;
  c0sb=110;
  c0lb=2190;
  c0ul=900;
  c0dl=37100;
  c0mb=3;
#
# ODE integration
  out=ode(func=model_1,times=tout,y=y0);
#
# Set up spline interpolation
  ppm0=200;dppm=100;max=16;
  splines(ppm0,dppm,max);
#
```

```
# Display selected output
  cat(sprintf("\n ncase = %4d\n",ncase));
  cat(sprintf(
    "\n    t      yla     yul     ppm     Tc      eps     pH
    r1\n"));
  for(it in 1:nout){
#
# CO2 emissions rate (in lower atmosphere)
  CO2_rate(tout[it],ncase);
  r1e[it]=r1;
#
# CO2 ppm (in lower atmosphere)
  ppm[it]=280*(1+out[it,2]);
#
# Total carbon
  Tc[it]=Tcs(ppm[it]);
#
# Evasion factor
  if(it==1){eps[1]=8.8;}
  if(it >1){
    num=(ppm[it]-ppm[it-1])/ppm[it-1];
    den=( Tc[it]- Tc[it-1])/ Tc[it-1];
    eps[it]=num/den;
  }
#
# Hydrogen ion
  hion[it]=hions(ppm[it]);
#
# pH
  pH[it]=pHs(ppm[it]);
#
# CO2 (CO2 + H2CO3 in upper layer)
  co2[it]=k0*ppm[it];
#
# Bicarbonate
  hco3[it]=k0*k1*ppm[it]/hion[it];
#
# Carbonate
  co3[it]=k0*k1*k2*ppm[it]/hion[it]^2;
#
# Boron
  boh4[it]=B/(1+hioni[it]/kb);
#
```

```
# Selected output
  cat(sprintf("%5.0f%8.4f%8.4f%8.1f%8.3f%8.3f%8.3f%8.4f\n",
    out[it,1],out[it,2],out[it,6],ppm[it],Tc[it],eps[it],
    pH[it],r1e[it]));
#
# CO2 GtC (in all reservoirs) vs t
  cla[it]=c0la*(1+out[it,2]);
  cua[it]=c0ua*(1+out[it,3]);
  csb[it]=c0sb*(1+out[it,4]);
  clb[it]=c0lb*(1+out[it,5]);
  cul[it]=c0ul*(1+out[it,6]);
  cdl[it]=c0dl*(1+out[it,7]);
  cmb[it]=c0mb*(1+out[it,8]);
  }
#
# CO2 ppm at 2013 (by linear interpolation between 2010
# and 2020)
  p2013=ppm[17]+(ppm[18]-ppm[17])*(2013-2010)/(2020-2010);
#
# GtC (in all reservoirs) at 2013
  cla2013=cla[17]+(cla[18]-cla[17])*(2013-2010)/(2020-2010);
  cua2013=cua[17]+(cua[18]-cua[17])*(2013-2010)/(2020-2010);
  clb2013=clb[17]+(clb[18]-clb[17])*(2013-2010)/(2020-2010);
  csb2013=csb[17]+(csb[18]-csb[17])*(2013-2010)/(2020-2010);
  cul2013=cul[17]+(cul[18]-cul[17])*(2013-2010)/(2020-2010);
  cdl2013=cdl[17]+(cdl[18]-cdl[17])*(2013-2010)/(2020-2010);
  cmb2013=cmb[17]+(cmb[18]-cmb[17])*(2013-2010)/(2020-2010);
#
# GtC at 2013 output
  cat(sprintf("\n\n Atmosphere ppm(2013) = %6.1f\n",p2013));
  cat(sprintf("\n GtC Lower Atmosphere(2013) = %6.1f\n",
    cla2013));
  cat(sprintf("\n GtC Upper Atmosphere(2013) = %6.1f\n",
    cua2013));
  cat(sprintf("\n GtC Long-lived(2013)       = %6.1f\n",
    clb2013));
  cat(sprintf("\n GtC Short-lived(2013)      = %6.1f\n",
    csb2013));
  cat(sprintf("\n GtC Upper Layer(2013)      = %6.1f\n",
    cul2013));
  cat(sprintf("\n GtC Deep Layer(2013)       = %6.1f\n",
    cdl2013));
```

```
  cat(sprintf("\n GtC Marine Biota(2013)     = %6.1f\n",
    cmb2013));
  cat(sprintf("\n ncall = %4d\n",ncall));
#
# CO2 sensitivity (C^o/ppm CO2)
  co2s=0.005;
#
# Select plots (0 - no plot, 1 - selected plot)
#
#   Default is no plots
    plot_out=rep(0,20);
#
#   Summary of available plots in plot_output.m
#
#   1 - c_{la}(ppm) vs t
#
#   2 - eps vs t
#
#   3 - pH vs t
#
#   4 - CO_2 + H_2CO_3 (millimols/liter) vs ppm
#
#   5 - HCO_3^- bicarbonate (millimols/liter) vs ppm
#
#   6 - CO_3^{2-} carbonate (millimols/liter) vs ppm
#
#   7 - B(OH)^-_4 (millimols/liter) vs ppm
#
#   8 - CO_2+H_2CO_3,HCO_3,CO_3 (fraction of 1850 values) vs
#       ppm
#
#   9 - y(1) (lower atmosphere), y(5) (ocean upper layer),
#       y(6) (ocean deep layer) (fractional change from 1850)
#       vs t
#
#  10 - co2s*(c_{la}(t) - c_{la}(1850)) (C^o/ppm*ppm = C^o)
#       vs t (to give an indication of warming)
#
#  11 - mass C vs t, cla, cua, csb, clb, cul, cdl, cmb (GtC)
#       vs t (semilog plot because of differences in magni-
#       tudes)
#
```

```
#  12 - composite plot of ppm CO2 in lower atmosphere vs t
#       (use only for four consecutive cases ncase=1,2,3,4)
#
#  Select plots
    plot_out[1]=1;
    plot_out[3]=1;
    plot_out[8]=1;
#
# Plot output
    plot_output(plot_out);
```

The main Matlab program is listed next.

Listing 1b: Main program in Matlab

```
%
% Clear previous files
  clear all
  clc
%
% General parameters (shared with other routines)
  global ncall ncase
%
% Spline tables
  global Tc hion pH
%
% Variables for plotting
  global t     y    ppm   co2   hco3   co3   boh4...
         cla   cua   csb   clb   cul   cdl   cmb...
         co2s  eps
%
% Equilibrium constants, B
  global k0 k1 k2 kb kw B
%
% Select case (see CO2_rate.m)
  ncase=1;
%
% Equilibrium constants (in millimols/lt units)
  k0=0.03347e-03;
  k1=9.747e-04;
  k2=8.501e-07;
  kb=1.881e-06;
  kw=6.46e-09;
```

```
%
% Boron
  B=0.409;
%
% Initial conditions
  n=7;
  y0=zeros(1,n);
%
% Independent variable for ODE integration
  t0=1850;tf=2100;nout=26;
  tout=[t0:10:tf]';
  ncall=0;
%
% Initialized carbon reservoirs (GtC)
  c0la=0.85*597;
  c0ua=0.15*597;
  c0sb=110;
  c0lb=2190;
  c0ul=900;
  c0dl=37100;
  c0mb=3;
%
% ODE integration
  reltol=1.0e-06;abstol=1.0e-06;
  options=odeset('RelTol',reltol,'AbsTol',abstol);
  [t,y]=ode15s(@model_1,tout,y0,options);
%
% Set up spline interpolation
  ppm0=200;dppm=100;max=16;
  [Tcs,hions,pHs]=splines(ppm0,dppm,max);
%
% Display selected output
  fprintf('\n ncase = %4d\n',ncase);
  fprintf(
  '\n    t      yla     yul     ppm     Tc      eps      pH
  r1\n')
  for it=1:nout
%
% CO2 emissions rate (in lower atmosphere)
  [c1,r1]=CO2_rate(t(it));
  r1_t(it)=r1;
%
```

```
% CO2 ppm (in lower atmosphere)
  ppm(it)=280*(1+y(it,1));
%
% Total carbon
  Tc(it)=ppval(Tcs,ppm(it));
%
% Evasion factor
  if(it==1)eps(1)=8.8;end
  if(it >1)
    num=(ppm(it)-ppm(it-1))/ppm(it-1);
    den=( Tc(it)- Tc(it-1))/ Tc(it-1);
    eps(it)=num/den;
  end
%
% Hydrogen ion
  hion(it)=ppval(hions,ppm(it));
%
% pH
  pH(it)=ppval(pHs,ppm(it));
%
% CO2 (CO2 + H2CO3 in upper layer)
  co2(it)=k0*ppm(it);
%
% Bicarbonate
  hco3(it)=k0*k1*ppm(it)/hion(it);
%
% Carbonate
  co3(it)=k0*k1*k2*ppm(it)/hion(it)^2;
%
% Boron
  boh4(it)=B/(1+hion(it)/kb);
%
% Selected output
  fprintf('%5.0f%8.4f%8.4f%8.1f%8.3f%8.3f%8.3f%8.4f\n',...
    t(it),y(it,1),y(it,5),ppm(it),Tc(it),eps(it),pH(it),
    r1_t(it));
%
% CO2 GtC (in all reservoirs) vs t
  cla(it)=c0la*(1+y(it,1));
  cua(it)=c0ua*(1+y(it,2));
  csb(it)=c0sb*(1+y(it,3));
  clb(it)=c0lb*(1+y(it,4));
```

```
    cul(it)=cOul*(1+y(it,5));
    cdl(it)=cOdl*(1+y(it,6));
    cmb(it)=cOmb*(1+y(it,7));
    end
%
% CO2 ppm at 2013 (by linear interpolation between 2010
% and 2020)
    p2013=ppm(17)+(ppm(18)-ppm(17))*(2013-2010)/(2020-2010);
%
% GtC (in all reservoirs) at 2013
    cla2013=cla(17)+(cla(18)-cla(17))*(2013-2010)/(2020-2010);
    cua2013=cua(17)+(cua(18)-cua(17))*(2013-2010)/(2020-2010);
    clb2013=clb(17)+(clb(18)-clb(17))*(2013-2010)/(2020-2010);
    csb2013=csb(17)+(csb(18)-csb(17))*(2013-2010)/(2020-2010);
    cul2013=cul(17)+(cul(18)-cul(17))*(2013-2010)/(2020-2010);
    cdl2013=cdl(17)+(cdl(18)-cdl(17))*(2013-2010)/(2020-2010);
    cmb2013=cmb(17)+(cmb(18)-cmb(17))*(2013-2010)/(2020-2010);
%
% GtC at 2013 output
    fprintf('\n\n Atmosphere ppm(2013) = %6.1f\n',p2013);
    fprintf('\n GtC Lower Atmosphere(2013) = %6.1f\n',cla2013);
    fprintf('\n GtC Upper Atmosphere(2013) = %6.1f\n',cua2013);
    fprintf('\n GtC Long-lived(2013)       = %6.1f\n',clb2013);
    fprintf('\n GtC Short-lived(2013)      = %6.1f\n',csb2013);
    fprintf('\n GtC Upper Layer(2013)      = %6.1f\n',cul2013);
    fprintf('\n GtC Deep Layer(2013)       = %6.1f\n',cdl2013);
    fprintf('\n GtC Marine Biota(2013)     = %6.1f\n',cmb2013);
    fprintf('\n ncall = %4d\n',ncall);
%
% CO2 sensitivity (C^o/ppm CO2)
    co2s=0.005;
%
% Select plots (0 - no plot, 1 - selected plot)
%
%   Default is no plots
    plot_out=zeros(20,1);
%
% Summary of available plots in plot_output.m
%
%   1 - c_{la}(ppm) vs t
%
```

```
%    2 - eps vs t
%
%    3 - pH vs t
%
%    4 - CO_2 + H_2CO_3 (millimols/liter) vs ppm
%
%    5 - HCO_3^- bicarbonate (millimols/liter) vs ppm
%
%    6 - CO_3^{2-} carbonate (millimols/liter) vs ppm
%
%    7 - B(OH)^-_4 (millimols/liter) vs ppm
%
%    8 - CO_2+H_2CO_3,HCO_3,CO_3 (fraction of 1850 values) vs
%        ppm
%
%    9 - y(1) (lower atmosphere), y(5) (ocean upper layer),
%        y(6) (ocean deep layer) (fractional change from 1850)
%        vs t
%
%   10 - co2s*(c_{la}(t) - c_{la}(1850)) (C^o/ppm*ppm = C^o)
%        vs t (to give an indication of warming)
%
%   11 - mass C vs t, cla, cua, csb, clb, cul, cdl, cmb (GtC)
%        vs t (semilog plot because of differences in magni-
%        tudes)
%
%   12 - composite plot of ppm CO2 in lower atmosphere vs t
%        (use only for four consecutive cases ncase=1,2,3,4)
     plot_out(1)=1;
     plot_out(3)=1;
     plot_out(8)=1;
%
% Plot output
  plot_output(plot_out);
```

We can note the following points about these main programs (Listings 1a and 1b):

- Previous R and Matlab files are cleared (the R and Matlab code can be easily distinguished by the comment identifiers # and %, respectively).

R:

```
#
# Delete previous workspaces
  rm(list=ls(all=TRUE))
```

Matlab:

```
%
% Clear previous files
  clear all
  clc
```

- Specific code used at the beginning of Listings 1a and 1b is explained.

R:

```
#
# Access ODE integrator
  library("deSolve");
#
# Access files
  setwd("c:/model_R");
  source("model_1.R");
  source("splines.R");
  source("CO2_rate.R");
  source("plot_output.R");
```

The R ODE library `deSolve` [26] and particular R files in the working directory `c:/model_R` are specified. The utility `source` is used to select the particular files (with extension R). A forward slash / is used rather than the usual backslash \ in the `setwd` (set working directory).

R: A special designation for shared variables and parameters is not required in R (variables and parameters in a R routine are shared with subordinate routines).

Matlab:

```
%
% General parameters (shared with other routines)
  global ncall ncase
```

```
%
% Spline tables
  global Tc hion pH
%
% Variables for plotting
  global t      y    ppm    co2   hco3    co3   boh4...
         cla   cua    csb   clb    cul    cdl    cmb...
         co2s  eps
%
% Equilibrium constants, B
  global k0 k1 k2 kb kw B
```

Variables and parameters are declared **global** so that they can be shared with other routines.

- Arrays are declared in R using the **rep** utility. **nout** is the number of output points in the computed solution.

R:

```
#
# Declare (preallocate) arrays
  nout=26;
    r1e=rep(0,nout);  ppm=rep(0,nout); eps=rep(0,nout);
     Tc=rep(0,nout); hion=rep(0,nout);  pH=rep(0,nout);
    co2=rep(0,nout); hco3=rep(0,nout); co3=rep(0,nout);
    boh4=rep(0,nout);
#
# Dimensional ODE dependent variables
  cla=rep(0,nout);cua=rep(0,nout);csb=rep(0,nout);
  clb=rep(0,nout);cul=rep(0,nout);cdl=rep(0,nout);
  cmb=rep(0,nout);
```

Matlab: Dimensioning of arrays is not required in Matlab (dynamic dimensioning takes place as the arrays are used).

- The case is selected from **ncase=1,2,3,4** as explained in functions **CO2_rate.R**, **CO2_rate.m** of Listings 4a and 4b.

R:

```
#
# Select case (see CO2_rate.R)
  ncase=1;
```

Matlab:

```
%
% Select case (see CO2_rate.m)
  ncase=1;
```

- The equilibrium constants used to calculate the ocean *pH* and the ocean boron concentration are defined numerically.

  R:

```
#
# Equilibrium constants (in millimols/lt units)
  k0=0.03347e-03;
  k1=9.747e-04;
  k2=8.501e-07;
  kb=1.881e-06;
  kw=6.46e-09;
#
# Boron
  B=0.409;
```

Matlab:

```
%
% Equilibrium constants (in millimols/lt units)
  k0=0.03347e-03;
  k1=9.747e-04;
  k2=8.501e-07;
  kb=1.881e-06;
  kw=6.46e-09;
%
% Boron
  B=0.409;
```

- Zero ICs (3.2), (3.3b) to (3.9b) are defined with the utilities **rep** and **zeros**. The model has n=7 ODEs.

  R:

```
#
# Initial conditions
  n=7;
  y0=rep(0,n);
```

Matlab:

```
%
% Initial conditions
  n=7;
  y0=zeros(1,n);
```

- The time scale $1850 \leq t \leq 2100$ is defined with nout=26 output points in the vector tout. Also, the counter for the calls to the ODE routines (discussed subsequently) is initialized.

R:

```
#
# Independent variable for ODE integration
  t0=1850;tf=2100;
  tout=seq(from=t0,to=tf,by=(tf-t0)/(nout-1));
  ncall=0;
```

Matlab:

```
%
% Independent variable for ODE integration
  t0=1850;tf=2100;nout=26;
  tout=[t0:10:tf]';
  ncall=0;
```

For R, the seq utility is used and the increment in the number of output points $((2100 - 1850)/(26 - 1) = 250/25 = 10)$ is computed. For Matlab, a vector definition is used ([::]) with a transpose (') since the ODE integrator requires a column vector for the IC vector.

- The initial ($t = 1850$) quantities of carbon in the seven reservoirs in GtC (Pg) are defined numerically.

R:

```
#
# Initialized carbon reservoirs (GtC)
  c0la=0.85*597;
  c0ua=0.15*597;
  c0sb=110;
  c0lb=2190;
  c0ul=900;
```

```
c0dl=37100;
c0mb=3;
```

Matlab:

```
%
% Initialized carbon reservoirs (GtC)
c0la=0.85*597;
c0ua=0.15*597;
c0sb=110;
c0lb=2190;
c0ul=900;
c0dl=37100;
c0mb=3;
```

The initial carbon in the upper and lower atmospheres (considered here to be the stratosphere and troposphere), 597 GtC, is divided as %15 and %85 [24], respectively. These inventories are used to convert the model dimensionless (relative) variables from Eqs. (3.3) to (3.9) to dimensional (absolute) values according to Eq. (3.1). For example, if Eq. (3.1) is rearranged as

$$c_{dim}(t) = c_{dim}(t = 1850) + c_{dim}(t = 1850)y(t)$$
$$= c_{dim}(t = 1850)(1 + y(t)), \qquad (5.1)$$

application of this equation to the lower atmosphere gives

$$c_{la}(t) = cola(1 + y_{la}(t)). \qquad (5.2)$$

In this way, the seven dimensional dependent variables can be computed from the dimensionless values as a function of t for tabulated and plotted output of the model, and for comparison with reported (literature) values. Note, however, that the calculations within the model (numerical integration of the seven ODEs) are in terms of the dimensionless variables.

Also, $c_{dim}(t = 1850)$ with any dimensional units can be used in the preceding calculation. For example,

$$c_{la}(t) = 280(1 + y_{la}(t)) \qquad (5.3)$$

is used below to calculate the lower atmospheric ppm CO_2 as a function of t where 280 is the ppm atmospheric CO_2 at 1850.[1]

- Equations (3.3) to (3.9) are integrated in R by the ODE integrator ode and in Matlab by ode15s.

R:

```
#
# ODE integration
  out=ode(func=model_1,times=tout,y=y0);
#
# Calls to ODE model_1
  cat(sprintf("\n ncall = %4d\n",ncall));
```

The 2D output from ODE integrator ode has a particular format. out[it,1] = t (with it = 1,2,...nout = 1,2,...26). out[it,2] to out[it,8] are the seven ODE dependent variables $y_1(t) = y_{ua}(t)$ to $y_7(t) = y_{mb}(t)$ (the solutions to Eqs. (3.3a) to (3.9a) as a function of t). Note that elements in arrays in R are defined with [ ] while elements in arrays in Matlab are defined with ( ).

Matlab:

```
%
% ODE integration
  reltol=1.0e-06;abstol=1.0e-06;
  options=odeset('RelTol',reltol,'AbsTol',abstol);
  [t,y]=ode15s(@model_1,tout,y0,options);
```

The 2D output from ODE integrator ode15s has the expected format. t in [t,y] is a vector of output values of t (the same values as in tout), and y in [t,y] is a 2D matrix with $y_1(t) = y_{ua}(t)$ to

---

[1]The atmospheric CO_2, in *ppm*, can be calculated from the total atmospheric carbon, *GtC*, as

$$ppm = \frac{(GtC/MWc)}{(Ma/MWa)}(10^6)$$

where

MWc	molecular weight of carbon = 12
MWa	molecular weight of air = 29
Ma	total mass of air in the atmosphere (Gt).

$y_7(t) = y_{mb}(t)$ as a function of t (the solutions to Eqs. (3.3a) to (3.9a) as a function of t).

model_1 is the routine that computes the RHS of Eqs. (3.3) to (3.9) (model_1.R in Listing 3a and model_1.m in Listing 3b).

- In the case of ode15s, error tolerances are specified so that they are the same as the default error tolerances of ode (the default of ode15s is 1.0e-03). In this way the two numerical solutions are brought into close agreement numerically as discussed below. Note for each integrator the use of the IC vector y0 which informs the integrator of the number of ODEs (by the length of y0, 7 in this case). Also, the vector of output points, tout, is an input to the integrator and the numerical solution is returned as a 2D array at these points (out for R and [t,y] for Matlab as explained previously). The display of these numerical solutions is discussed subsequently.

- A call to routines splines.m and splines.R sets up the spline interpolation of some quantities computed algebraically from the seven dependent variables (dimensionless concentrations) of Eqs. (3.3) to (3.9). The details of the spline routines (in Listings 5a and 5b) are discussed subsequently.

R:

```
#
# Set up spline interpolation
  ppm0=200;dppm=100;max=16;
  splines(ppm0,dppm,max);
```

Matlab:

```
%
% Set up spline interpolation
  ppm0=200;dppm=100;max=16;
  [Tcs,hions,pHs]=splines(ppm0,dppm,max);
```

The calls to splines set up tables of model variables based on the atmospheric CO_2 concentration ranging from 200 ppm to $200 + (16 - 1)(100) = 1700$. These tables are discussed next.

- Selected numerical output is produced by stepping through the nout = 26 values of t with a for using index it (with some code indented and placed on two lines so that it fits on a page).

R:

```
#
# Display selected output
  cat(sprintf("\n ncase = %4d\n",ncase));
  cat(sprintf(
  "\n   t      yla    yul    ppm    Tc     eps      pH
  r1\n"));
  for(it in 1:nout){
#
# CO2 emissions rate (in lower atmosphere)
  CO2_rate(tout[it],ncase);
  r1e[it]=r1;
#
# CO2 ppm (in lower atmosphere)
  ppm[it]=280*(1+out[it,2]);
#
# Total carbon
  Tc[it]=Tcs(ppm[it]);
#
# Evasion factor
  if(it==1){upu[1] 0.0;}
  if(it >1){
    num=(ppm[it]-ppm[it-1])/ppm[it-1];
    den=( Tc[it]- Tc[it-1])/ Tc[it-1];
    eps[it]=num/den;
  }
#
# Hydrogen ion
  hion[it]=hions(ppm[it]);
#
# pH
  pH[it]=pHs(ppm[it]);
#
# CO2 (CO2 + H2CO3 in upper layer)
  co2[it]=k0*ppm[it];
#
# Bicarbonate
  hco3[it]=k0*k1*ppm[it]/hion[it];
#
# Carbonate
  co3[it]=k0*k1*k2*ppm[it]/hion[it]^2;
#
```

```
# Boron
  boh4[it]=B/(1+hioni[it]/kb);
#
# Selected output
  cat(sprintf("%5.0f%8.4f%8.4f%8.1f%8.3f%8.3f%8.3f%8.4f\n",
    out[it,1],out[it,2],out[it,6],ppm[it],Tc[it],eps[it],
    pH[it],r1e[it]));
```

Matlab:

```
%
% Display selected output
  fprintf('\n ncase = %4d\n',ncase);
  fprintf(
  '\n    t      yla     yul     ppm      Tc      eps      pH
  r1\n')
  for it=1:nout
%
% CO2 emissions rate (in lower atmosphere)
  [c1,r1]=CO2_rate(t(it));
  r1_t(it)=r1;
%
% CO2 ppm (in lower atmosphere)
  ppm(it)=280*(1+y(it,1));
%
% Total carbon
  Tc(it)=ppval(Tcs,ppm(it));
%
% Evasion factor
  if(it==1)eps(1)=8.8;end
  if(it >1)
    num=(ppm(it)-ppm(it-1))/ppm(it-1);
    den=( Tc(it)- Tc(it-1))/ Tc(it-1);
    eps(it)=num/den;
  end
%
% Hydrogen ion
  hion(it)=ppval(hions,ppm(it));
%
% pH
  pH(it)=ppval(pHs,ppm(it));
%
```

```
% CO2 (CO2 + H2CO3 in upper layer)
  co2(it)=k0*ppm(it);
%
% Bicarbonate
  hco3(it)=k0*k1*ppm(it)/hion(it);
%
% Carbonate
  co3(it)=k0*k1*k2*ppm(it)/hion(it)^2;
%
% Boron
  boh4(it)=B/(1+hion(it)/kb);
%
% Selected output
  fprintf('%5.0f%8.4f%8.4f%8.1f%8.3f%8.3f%8.3f%8.4f\n',...
      t(it),y(it,1),y(it,5),ppm(it),Tc(it),eps(it),pH(it),
      r1_t(it));
```

We can note the following additional details of the **for** with index
it.

(n) The numerical solution is labeled with

R:

```
#
# Display selected output
  cat(sprintf("\n ncase = %4d\n",ncase));
  cat(sprintf(
  "\n   t   yla   yul   ppm    Tc    eps    pH
  r1\n"));
```

Matlab:

```
%
% Display selected output
  fprintf('\n ncase = %4d\n',ncase);
  fprintf(
  '\n   t   yla   yul   ppm    Tc    eps    pH
  r1\n')
```

The output statement for R is **cat(sprintf())** and for Matlab it is **fprintf()**. pH is included in the output since it is of
particular interest in ocean acidification [9].

(b) Within the `for`, the constants c_1, r_1 of Eq. (3.10) are first evaluated by a call to functions `CO2_rate.R`, `CO2_rate.m` (Listings 4a and 4b). For some of the cases discussed subsequently, r_1 varies with t, so r_1 is included in the output (the last item labeled above).

R:

```
   for(it in 1:nout){
#
# CO2 emissions rate (in lower atmosphere)
   CO2_rate(tout[it],ncase);
   r1e[it]=r1;
```

Matlab:

```
   for it=1:nout
%
% CO2 emissions rate (in lower atmosphere)
   [c1,r1]=CO2_rate(t(it));
   r1_t(it)=r1;
```

(c) Equation (5.3) is used to calculate the atmospheric CO_2 concentration `ppm[it]`, `ppm(it)` as a function of t (with index `it`). `ppm[it]`, `ppm(it)` are then used to calculate additional variables.

R:

```
#
# CO2 ppm (in lower atmosphere)
   ppm[it]=280*(1+out[it,2]);
#
```

Matlab:

```
%
% CO2 ppm (in lower atmosphere)
   ppm(it)=280*(1+y(it,1));
```

(d) Variables `Tc,hion,pH` for R and for Matlab pertaining to the carbon chemistry of the ocean are calculated through spline interpolation (the spline functions `Tcs,hions,pHs` are set up by the previous calls to `splines.R` and `splines.m`, Listings 5a and 5b).

R:

```
#
# Total carbon
  Tc[it]=Tcs(ppm[it]);
#
# Evasion factor
  if(it==1){eps[1]=8.8;}
  if(it >1){
    num=(ppm[it]-ppm[it-1])/ppm[it-1];
    den=( Tc[it]- Tc[it-1])/ Tc[it-1];
    eps[it]=num/den;
  }
#
# Hydrogen ion
  hion[it]=hions(ppm[it]);
#
# pH
  pH[it]=pHs(ppm[it]);
```

Matlab:

```
%
% Total carbon
  Tc(it)=ppval(Tcs,ppm(it));
%
% Evasion factor
  if(it==1)eps(1)=8.8;end
  if(it >1)
    num=(ppm(it)-ppm(it-1))/ppm(it-1);
    den=( Tc(it)- Tc(it-1))/ Tc(it-1);
    eps(it)=num/don;
  end
%
% Hydrogen ion
  hion(it)=ppval(hions,ppm(it));
%
% pH
  pH(it)=ppval(pHs,ppm(it));
```

Note the difference between R and Matlab for using the spline functions. ppval is the Matlab utility for interpolation using spline functions.

The evasion factor `eps` is calculated from the definition

$$\varepsilon = \frac{\delta ppm/ppm}{\delta Tc/Tc},$$

$\delta ppm = \text{ppm(it)} - \text{ppm(it} - 1)$ and $\delta Tc = \text{Tc(it)} - \text{Tc(it} - 1)$ (Matlab) are increments in ppm and Tc based on the 10-year intervals indexed as `it-1` and `it` (set in the `for` of `it`).

(e) Additional variables (`CO2 + H2CO3`), `bicarbonate`, `carbonate`, `boron` are calculated through equilibrium constants, e.g., `k0,k1,k2`. The various calculated variables are identified by comments and the calculations are not explained here; rather, the ocean carbon chemistry is discussed subsequently.

• The seven dimensionless dependent variables returned by the ODE integrator `ode` or `ode15s` are converted to dimensional reservoir inventories according to Eq. (5.1) based on the initial reservoir inventories `c0la` to `c0mb` from [24].

R:

```
#
# CO2 GtC (in all reservoirs) vs t
  cla[it]=c0la*(1+out[it,2]);
  cua[it]=c0ua*(1+out[it,3]);
  csb[it]=c0sb*(1+out[it,4]);
  clb[it]=c0lb*(1+out[it,5]);
  cul[it]=c0ul*(1+out[it,6]);
  cdl[it]=c0dl*(1+out[it,7]);
  cmb[it]=c0mb*(1+out[it,8]);
}
```

Matlab:

```
%
% CO2 GtC (in all reservoirs) vs t
  cla(it)=c0la*(1+y(it,1));
  cua(it)=c0ua*(1+y(it,2));
  csb(it)=c0sb*(1+y(it,3));
  clb(it)=c0lb*(1+y(it,4));
  cul(it)=c0ul*(1+y(it,5));
```

```
cdl(it)=c0dl*(1+y(it,6));
cmb(it)=c0mb*(1+y(it,7));
end
```

The final } and **end** terminate the **for** that steps the calculations and output through the **nout** = 26 values of t (with index **it**).

- The value of $c_{la}(2013)$ in ppm is of particular interest since this was reported to have passed 400 (*New York Times* [12]). It is estimated by linear interpolation between $c_{la}(2010)$ = **ppm(17)** and $c_{la}(2020)$ = **ppm(18)**. The value $c_{la}(2013) = 400$ is used as a reference point in comparing the numerical model output with the observed CO_2 at $t = 2013$ (c_1, r_1 in Eq. (3.10) are selected to produce this value as explained subsequently).

R:

```
#
# CO2 ppm at 2013 (by linear interpolation between 2010
# and 2020)
  p2013=ppm[17]+(ppm[18]-ppm[17])*(2013-2010)/(2020-2010);
```

Matlab:

```
%
% CO2 ppm at 2013 (by linear interpolation between 2010
% and 2020)
  p2013=ppm(17)+(ppm(18)-ppm(17))*(2013-2010)/(2020-2010);
```

- The seven reservoir inventories (in GtC or Pg) are computed and displayed for 2013 to complement the reported value **ppm(t=2013)=p2013=400** [12].

R:

```
#
# GtC (in all reservoirs) at 2013
  cla2013=cla[17]+(cla[18]-cla[17])*(2013-2010)/(2020-2010);
  cua2013=cua[17]+(cua[18]-cua[17])*(2013-2010)/(2020-2010);
  clb2013=clb[17]+(clb[18]-clb[17])*(2013-2010)/(2020-2010);
  csb2013=csb[17]+(csb[18]-csb[17])*(2013-2010)/(2020-2010);
  cul2013=cul[17]+(cul[18]-cul[17])*(2013-2010)/(2020-2010);
  cdl2013=cdl[17]+(cdl[18]-cdl[17])*(2013-2010)/(2020-2010);
  cmb2013=cmb[17]+(cmb[18]-cmb[17])*(2013-2010)/(2020-2010);
```

```
#
# GtC at 2013 output
  cat(sprintf("\n\n Atmosphere ppm(2013) = %6.1f\n",p2013));
  cat(sprintf("\n GtC Lower Atmosphere(2013) = %6.1f\n",
    cla2013));
  cat(sprintf("\n GtC Upper Atmosphere(2013) = %6.1f\n",
    cua2013));
  cat(sprintf("\n GtC Long-lived(2013)       = %6.1f\n",
    clb2013));
  cat(sprintf("\n GtC Short-lived(2013)      = %6.1f\n",
    csb2013));
  cat(sprintf("\n GtC Upper Layer(2013)      = %6.1f\n",
    cul2013));
  cat(sprintf("\n GtC Deep Layer(2013)       = %6.1f\n",
    cdl2013));
  cat(sprintf("\n GtC Marine Biota(2013)     = %6.1f\n",
    cmb2013));
  cat(sprintf("\n ncall = %4d\n",ncall));
#
# CO2 sensitivity (C^o/ppm CO2)
  co2s=0.005;
```

Matlab:

```
%
% GtC (in all reservoirs) at 2013
  cla2013=cla(17)+(cla(18)-cla(17))*(2013-2010)/(2020-2010);
  cua2013=cua(17)+(cua(18)-cua(17))*(2013-2010)/(2020-2010);
  clb2013=clb(17)+(clb(18)-clb(17))*(2013-2010)/(2020-2010);
  csb2013=csb(17)+(csb(18)-csb(17))*(2013-2010)/(2020-2010);
  cul2013=cul(17)+(cul(18)-cul(17))*(2013-2010)/(2020-2010);
  cdl2013=cdl(17)+(cdl(18)-cdl(17))*(2013-2010)/(2020-2010);
  cmb2013=cmb(17)+(cmb(18)-cmb(17))*(2013-2010)/(2020-2010);
%
% GtC at 2013 output
  fprintf('\n\n Atmosphere ppm(2013) = %6.1f\n',p2013);
  fprintf('\n GtC Lower Atmosphere(2013) = %6.1f\n',cla2013);
  fprintf('\n GtC Upper Atmosphere(2013) = %6.1f\n',cua2013);
  fprintf('\n GtC Long-lived(2013)       = %6.1f\n',clb2013);
  fprintf('\n GtC Short-lived(2013)      = %6.1f\n',csb2013);
  fprintf('\n GtC Upper Layer(2013)      = %6.1f\n',cul2013);
  fprintf('\n GtC Deep Layer(2013)       = %6.1f\n',cdl2013);
```

```
      fprintf('\n GtC Marine Biota(2013)       = %6.1f\n',cmb2013);
      fprintf('\n ncall = %4d\n',ncall);
%
% CO2 sensitivity (C^o/ppm CO2)
      co2s=0.005;
```

The number of calls to the ODE routine `model_1`, `ncall`, is displayed after the solutions are computed as a measure of the computational effort required to calculate the solutions.

A temperature sensitivity, `co2s`, is also defined to provide an estimate of global warming in the subsequent plotting. This sensitivity is $0.005°C/ppm$ so that an increase of 120 ppm (e.g., from $ppm(1850) = 280$ to $ppm(2013) = 400$) in atmospheric CO_2 produces a temperature rise of $(0.005)(120) = 0.6°C = (0.6)(1.8) = 1.08°F$.

- A series of plots is then produced by a call to `plot_output.R`, `plot_output.m` (Listings 2a and 2b). The particular set of plots selected by array `plot_out.R, plot_out.m` (initially zero) is 1,3,8. These three plots are discussed subsequently.

R:

```
#
# Select plots (0 - no plot, 1 - selected plot)
#
#    Default is no plots
     plot_out=rep(0,20);
#
#    Summary of available plots in plot_output.m
#
#    1 - c_{la}(ppm) vs t
#
#    2 - eps vs t
#
#    3 - pH vs t
#
#    4 - CO_2 + H_2CO_3 (millimols/liter) vs ppm
#
#    5 - HCO_3^- bicarbonate (millimols/liter) vs ppm
#
#    6 - CO_3^{2-} carbonate (millimols/liter) vs ppm
```

```
#
#   7 - B(OH)^-_4 (millimols/liter) vs ppm
#
#   8 - CO_2+H_2CO_3,HCO_3,CO_3 (fraction of 1850 values) vs
#       ppm
#
#   9 - y(1) (lower atmosphere), y(5) (ocean upper layer),
#       y(6) (ocean deep layer) (fractional change from 1850)
#       vs t
#
#  10 - co2s*(c_{la}(t) - c_{la}(1850)) (C^o/ppm*ppm = C^o)
#       vs t (to give an indication of warming)
#
#  11 - mass C vs t, cla, cua, csb, clb, cul, cdl, cmb (GtC)
#       vs t (semilog plot because of differences in magni-
#       tudes)
#
#  12 - composite plot of ppm CO2 in lower atmosphere vs t
#       (use only for four consecutive cases ncase=1,2,3,4)
#
#   Select plots
     plot_out[1]=1;
     plot_out[3]=1;
     plot_out[8]=1;
#
# Plot output
   plot_output(plot_out);
```

Matlab:

```
%
% Select plots (0 - no plot, 1 - selected plot)
%
%   Default is no plots
     plot_out=zeros(20,1);
%
% Summary of available plots in plot_output.m
%
%   1 - c_{la}(ppm) vs t
%
%   2 - eps vs t
%
```

```
%   3 - pH vs t
%
%   4 - CO_2 + H_2CO_3 (millimols/liter) vs ppm
%
%   5 - HCO_3^- bicarbonate (millimols/liter) vs ppm
%
%   6 - CO_3^{2-} carbonate (millimols/liter) vs ppm
%
%   7 - B(OH)^-_4 (millimols/liter) vs ppm
%
%   8 - CO_2+H_2CO_3,HCO_3,CO_3 (fraction of 1850 values)
%
%   9 - y(1) (lower atmosphere), y(5) (ocean upper layer),
%       y(6) (ocean deep layer) (fractional change from 1850)
%       vs t
%
%  10 - co2s*(c_{la}(t) - c_{la}(1850)) (C^o/ppm*ppm = C^o)
%       vs t (to give an indication of warming)
%
%  11 - mass C vs t, cla, cua, csb, clb, cul, cdl, cmb (GtC)
%       vs t (semilog plot because of differences in magni-
%       tudes)
%
%  12 - composite plot of ppm CO2 in lower atmosphere vs t
%       (use only for four consecutive cases ncase=1,2,3,4)
     plot_out(1)=1;
     plot_out(3)=1;
     plot_out(8)=1;
%
% Plot output
     plot_output(plot_out);
```

plot_out[12]=1, plot_out(12)=1 produces a composite plot for four cases (ncase=1,2,3,4) as explained subsequently.

This completes the discussion of the main programs in Listings 1a and 1b. We now consider the plotting routines plot_output.R, plot_output.m.

5.2. Plotting Routines

The plotting routines plot_output.R, plot_output.m called by the main programs of Listings 1a and 1b are listed next.

Listing 2a: Routine plot_output.R for plot_out[1] to plot_out[12]

R:

```
  plot_output=function(plot_out){
#
# Select plots
#
# c_{la}(ppm) vs t
  if(plot_out[1]==1){
  par(mfrow=c(1,1))
  plot(tout,ppm,xlab="t",ylab="c_{la}(ppm)",type="l",lwd=2,
      main="c_{la}(ppm) vs t");
  }
#
# eps vs t
  if(plot_out[2]==1){
  par(mfrow=c(1,1))
  plot(tout,eps,xlab="t",ylab="eps(t)",type="l",lwd=2,
      main="eps(t) vs t");
  }
#
# pH vs t
  if(plot_out[3]==1){
  par(mfrow=c(1,1))
  plot(tout,pH,xlab="t",ylab="pH(t)",type="l",lwd=2,
      main="pH(t) vs t");
  }
#
# CO_2 + H_2CO_3 (millimols/liter) vs ppm
  if(plot_out[4]==1){
  par(mfrow=c(1,1))
  plot(ppm,co2,xlab="c_{la} (ppm)",ylab="CO_2 + H_2CO_3",
      type="l",lwd=2,main="CO_2 + H_2CO_3 (millimols/liter)
      vs c_{la} (ppm)");
  }
```

```
#
# HCO_3^{-} - bicarbonate (millimols/liter) vs ppm
  if(plot_out[5]==1){
  par(mfrow=c(1,1))
  plot(ppm,hco3,xlab="c_{la} (ppm)",ylab="HCO_3^{-}",type="l",
      lwd=2,main="HCO_3^{-} - bicarbonate (millimols/liter)
      vs c_{la} (ppm)");
  }
#
# CO_3^{2-} - carbonate (millimols/liter) vs ppm
  if(plot_out[6]==1){
  par(mfrow=c(1,1))
  plot(ppm,co3,xlab="c_{la} (ppm)",ylab="CO_3^{2-}",type="l",
      lwd=2,main="CO_3^{2-} - carbonate (millimols/liter)
      vs c_{la} (ppm)");
  }
#
# B(OH)^{-}_4 (millimols/liter) vs ppm
  if(plot_out[7]==1){
  par(mfrow=c(1,1))
  plot(ppm,boh4,xlab="c_{la} (ppm)",ylab="B(OH)^{-}_4}",
      type="l",lwd=2,main="B(OH)^{-}_4 (millimols/liter)
      vs c_{la} (ppm)");
  }
#
# CO_2+H_2CO_3,HCO_3^{-},CO_3^{2-} (fractional change from 1850)
# vs ppm
  if(plot_out[8]==1){
  par(mfrow=c(1,1))
  plot(ppm,co2/co2[1],xlab="ppm",ylab="c(frac)",type="l",lwd=2,
      main="Fractional upper layer\n
            co2+h2co3,hco3,co3 vs cla (ppm)",ylim=c(0,3.5));
  points(ppm,co2/co2[1],pch="1",lwd=2);
   lines(ppm,hco3/hco3[1],type="l",lwd=2);
  points(ppm,hco3/hco3[1], pch="2",lwd=2);
   lines(ppm,co3/co3[1],type="l",lwd=2);
  points(ppm,co3/co3[1], pch="3",lwd=2);
  }
#
# y(1) (lower atmosphere), y(5) (ocean upper layer), y(6)
# (ocean deep layer) (fractional change from 1850) vs t
  if(plot_out[9]==1){
```

```
   par(mfrow=c(1,1))
   plot(tout,cla/cla[1],xlab="t",ylab="y(frac)",type="l",lwd=2,
       main="Fractional cla,cul,cdl vs t",pch="1");
   points(tout,cla/cla[1], pch="1",lwd=2);
    lines(tout,cul/cul[1],type="l",lwd=2);
   points(tout,cul/cul[1], pch="2",lwd=2);
    lines(tout,cdl/cdl[1],type="l",lwd=2);
   points(tout,cdl/cdl[1], pch="3",lwd=2);
   }
#
# co2s*(c_{la}(t) - c_{la}(1850)) (C^o/ppm*ppm = C^o) vs t
# (to give an indication of warming)
   if(plot_out[10]==1){
   par(mfrow=c(1,1))
   plot(tout,co2s*(ppm-ppm[1]),xlab="t",ylab="warming (C^o)",
       type="l",lwd=2,main="co2s*(cla(t)-cla(1850)) vs t");
   }
#
# Mass C vs t, cla, cua, csb, clb, cul, cdl, cmb (GtC)
# vs t (semilog plot because of differences in magnitudes)
   if(plot_out[11]==1){
   par(mfrow=c(1,1))
   plot(tout,log10(cla),xlab="t",ylab="log10(mass CO2 (GtC))",
       type="l",lwd=2,main="log10 of cla,cua,csb,ckb,cul,cdl,
       cmb vs t",pch="1",ylim=c(0,5));
   points(tout,log10(cla), pch="1",lwd=2);
    lines(tout,log10(cua),type="l",lwd=2);
   points(tout,log10(cua), pch="2",lwd=2);
    lines(tout,log10(csb),type="l",lwd=2);
   points(tout,log10(csb), pch="3",lwd=2);
    lines(tout,log10(clb),type="l",lwd=2);
   points(tout,log10(clb), pch="4",lwd=2);
    lines(tout,log10(cul),type="l",lwd=2);
   points(tout,log10(cul), pch="5",lwd=2);
    lines(tout,log10(cdl),type="l",lwd=2);
   points(tout,log10(cdl), pch="6",lwd=2);
    lines(tout,log10(cmb),type="l",lwd=2);
   points(tout,log10(cmb), pch="7",lwd=2);
   }
#
# Composite plot of ppm CO2 in lower atmosphere vs t
# (use only for four consecutive cases ncase=1,2,3,4)
```

```
if(plot_out[12]==1){
par(mfrow=c(1,1))
plot(tout,ppm_case[,1],xlab="t",ylab="cla (ppm)",
     type="l",lwd=2,main="cla (ppm) vs t,
     \n variation in rate of CO2 emissions",pch="1");
points(tout,ppm_case[,1], pch="1",lwd=2);
 lines(tout,ppm_case[,2],type="l",lwd=2);
points(tout,ppm_case[,2], pch="2",lwd=2);
 lines(tout,ppm_case[,3],type="l",lwd=2);
points(tout,ppm_case[,3], pch="3",lwd=2);
 lines(tout,ppm_case[,4],type="l",lwd=2);
points(tout,ppm_case[,4], pch="4",lwd=2);
 }
#
# End of plot_output
 }
```

Listing 2b: Routine plot_output.m for plot_out(1) to plot_out(12)

Matlab:

```
function plot_output(plot_out)
%
% General parameters
  global ncall ncase
%
% Spline tables
  global hion Tc pH
%
% Variables for plotting
  global t      y    ppm   eps    pH   co2   hco3   co3  boh4...
               cla   cua   csb    clb  cul   cdl    cmb  co2s...
        ppm_case
%
% Counter for plot
  nplot=0;
%
% Select plots
%
% c_[la](ppm) vs t
```

```
   if plot_out(1)==1
   nplot=nplot+1;
   figure(nplot);
   plot(t,ppm,'-');
   title('c_{la}(ppm) vs t'); xlabel('t'); ylabel('c_{la}
         (ppm)');
   end
%
% eps vs t
   if plot_out(2)==1
   nplot=nplot+1;
   figure(nplot);
   plot(t,eps,'-');
   title('eps vs t'); xlabel('t'); ylabel('eps');
   end
%
% pH vs t
   if plot_out(3)==1
   nplot=nplot+1;
   figure(nplot);
   plot(t,pH,'-');
   title('pH vs t'); xlabel('t'); ylabel('pH');
   end
%
% CO_2 + H_2CO_3 (millimols/liter) vs ppm
   if plot_out(4)==1
   nplot=nplot+1;
   figure(nplot);
   plot(ppm,co2,'-')
   axis tight
   title('CO_2 + H_2CO_3 (millimols/liter) vs c_{la} (ppm)');
   xlabel('c_{la} (ppm)'); ylabel('CO_2 + H_2CO_3');
   end
%
% HCO_3^- - bicarbonate (millimols/liter) vs ppm
   if plot_out(5)==1
   nplot=nplot+1;
   figure(nplot);
   plot(ppm,hco3,'-');
   axis tight
   title('HCO_3^- - bicarbonate (millimols/liter) vs c_{la}
         (ppm)');
```

```
  xlabel('c_{la} (ppm)'); ylabel('HCO_3^-');
  end
%
% CO_3^{2-} - carbonate (millimols/liter) vs ppm
  if plot_out(6)==1
  nplot=nplot+1;
  figure(nplot);
  plot(ppm,co3,'-');
  axis tight
  title('CO_3^{2-} - carbonate (millimols/liter) vs c_{la}
       (ppm)');
  xlabel('c_{la} (ppm)'); ylabel('CO_3^{2-}');
  end
%
% B(OH)^-_4 (millimols/liter) vs ppm
  if plot_out(7)==1
  nplot=nplot+1;
  figure(nplot);
  plot(ppm,boh4,'-');
  axis tight
  title('B(OH)^- 4 (millimols/liter) vs c_{la} (ppm)');
  xlabel('c_{la} (ppm)'); ylabel('B(OH)^-_4');
  end
%
% CO_2+H2CO3,HCO_3,CO_3  (fractional change from 1850) vs ppm
  if plot_out(8)==1
  nplot=nplot+1;
  figure(nplot);
  plot(ppm,co2/co2(1),'o-',ppm,hco3/hco3(1),'+-',ppm,
       co3/co3(1),'x-');
  axis tight
  title('Fractional upper layer co2+h2co3,hco3,co3 vs cla
       (ppm)');
  xlabel('ppm'); ylabel('c(frac)');
  legend('CO_2+H_2CO_3','HCO_3','CO_3','Location','NorthWest');
  end
%
% y(1) (lower atmosphere), y(5) (ocean upper layer), y(6)
% (ocean deep layer) (fractional change from 1850) vs t
  if plot_out(9)==1
  nplot=nplot+1;
  figure(nplot);
```

```
plot(t,y(:,1),'o-',t,y(:,5),'+-',t,y(:,6),'x-');
title('Fractional cla,cul,cdl vs t'); xlabel('t');
        ylabel('y(frac)')
legend('lower atmosphere','upper layer','deep layer',...
        'Location','NorthWest');
end
%
% co2s*(c_{la}(t) - c_{la}(1850)) (C^o/ppm*ppm = C^o) vs t
% (to give an indication of warming)
if plot_out(10)==1
nplot=nplot+1;
figure(nplot);
plot(t,co2s*(ppm-ppm(1)),'-');
title('co2s*(cla(t)-cla(1850)) vs t');
xlabel('t'); ylabel('warming (C^o)');
end
%
% Mass C vs t, cla, cua, csb, clb, cul, cdl, cmb (GtC)
% vs t (semilog plot because of differences in magnitudes)
if plot_out(11)==1
nplot=nplot+1;
figure(nplot);
semilogy(t,cla,'o-',t,cua,'+-',t,csb,'x-',t,clb,'v-',
        t,cul,'x-',t,cdl,'+-',t,cmb,'o-');
title('log10 of cla,cua,csb,ckb,cul,cdl,cmb vs t');
xlabel('t'); ylabel('log(mass CO2 (GtC))');
legend('lower atmosphere','upper atmosphere','short-lived
        biota','long-lived biota','upper ocean layer',
        'deep ocean layer','marine biosphere','Location',
        'NorthWest');
end
%
% Composite plot of ppm CO2 in lower atmosphere vs t
% (use only for four consecutive cases ncase=1,2,3,4)
if plot_out(12)==1
nplot=nplot+1;
figure(nplot)
plot(t,ppm_case(:,1),'o-',t,ppm_case(:,2),'x-',t,
        ppm_case(:,3),'+-',t,ppm_case(:,4),'*-');
axis([1850 2100 250 900]);
legend('rc1 = 0.0050','rc1 = 0.0025','rc1 = 0',
        'rc1 = -0.0100','Location','NorthWest');
```

```
title(' cla (ppm) vs t; variation in rate of CO2
     emissions');
xlabel('t'); ylabel('cla (ppm)');
end
%
% End of plot_output
end
```

We can note the following points about plot_output.R, plot_output.m.

- The routines start with a defining statement. plot_output.m has a global area with the arrays to be plotted. plot_output.R receives these arrays by default (without a special designation such as global).

R:

```
plot_output=function(plot_out){
```

Matlab:

```
function plot_output(plot_out)
%
% General parameters
  global ncall ncase
%
% Spline tables
  global hion Tc pH
%
% Variables for plotting
  global t      y    ppm   eps   pH   co2  hco3   co3  boh4...
               cla   cua   csb   clb   cul  cdl   cmb  co2s...
           ppm_case
%
% Counter for plot
  nplot=0;
```

For plot_output.m, a counter for the plots, nplot, is initialized at the beginning. This counter is then used to number the plots with figure(nplot) (discussed next).

- The routines are essentially a set of repeated statements for `plot_out[1]` to `plot_out[12]` and `plot_out(1)` to `plot_out(12)`. For example, for the first plot.

R:

```
#
# Select plots
#
# c_{la}(ppm) vs t
  if(plot_out[1]==1){
  par(mfrow=c(1,1))
  plot(tout,ppm,xlab="t",ylab="ppm(t)",type="l",lwd=2,
      main="ppm(t) vs t");
  }
```

The basic plotting utility `plot` with labels `xlab`, `ylab`, `main` is essentially self-explanatory and is not discussed here; also the resulting plot is discussed subsequently (which has a figure number Fig. 2a). Note that for no plot, `plot_out[1]` = 0, while for a plot, `plot_out[1]` = 1 (as set in the calling program of Listing 1a).

Matlab:

```
%
% Select plots
%
% c_{la}(ppm) vs t
  if plot_out(1)==1
  nplot=nplot+1;
  figure(nplot);
  plot(t,ppm,'-');
  title('c_{la}(ppm) vs t'); xlabel('t'); ylabel('c_{la}
      (ppm)');
  end
```

The basic plotting utilities `figure, plot, title, xlabel, ylabel` are essentially self-explanatory and are not discussed here; also the resulting plot is discussed subsequently (which has a figure number of 1 from the statement `figure(nplot)` and also Fig. 2b in the discussion). Note that for no plot, `plot_out(1)` = 0, while for a plot, `plot_out(1)` = 1 (as set in the calling program of Listing 1b).

- This basic set of plotting statements is repeated through 12 cases (refer to Listings 2a and 2b). For the main programs in Listings 1a and 1b, the selected plots are 1,3,8.

Additional plots can easily be added to `plot_output.R`, `plot_output.m` by repeating the basic set of plotting statements embedded in an `if`.

The ODE routines, `model_1.R`, `model_1.m` called by `ode`, `ode15s` are considered next.

5.3. ODE Routines

The ODE functions `model_1.R`, `model_1.m` called by ODE integrators `ode`, `ode15s` in the main programs of Listings 1a and 1b are listed next. These routines have the programming of Eqs. (3.3) to (3.10).

Listing 3a: Function `model_1.R` for Eqs. (3.3) to (3.10)

R:

```
  model_1=function(t,y,parms){
#
# Function model_1 computes the temporal derivatives
# of the seven dependent variables
#
# Model dependent variables
  yla=y[1];
  yua=y[2];
  ysb=y[3];
  ylb=y[4];
  yul=y[5];
  ydl=y[6];
  ymb=y[7];
#
# ODEs
  CO2_rate(t,ncase);
  dyla-    1/5*(yua-yla)+
           1/0.75*(ysb-yla)+
```

```
                1/150*(ylb-yla)+
                1/30*(yul-yla)+
                c1*exp(r1*(t-1850)));
    dyua=       1/3*(yla-yua);
    dysb=    1/0.75*(yla-ysb);
    dylb=    1/150*(yla-ylb);
    dyul=     1/80*(yla-yul)+
              1/200*(ydl-yul)+
                1/5*(ymb-yul);
    dydl=  1/1500*(yul-ydl);
    dymb=     1/10*(yul-ymb);
#
# Derivative vector
  yt=rep(0,7);
  yt[1]=dyla;
  yt[2]=dyua;
  yt[3]=dysb;
  yt[4]=dylb;
  yt[5]=dyul;
  yt[6]=dydl;
  yt[7]=dymb;
#
# Increment calls to model_1
  ncall <<- ncall+1;
#
# Return derivative vector
  return(list(c(yt)));
#
# End of model_1
  }
```

Listing 3b: Function `model_1.m` for Eqs. (3.3) to (3.10)

Matlab:

```
  function yt=model_1(t,y)
%
% Function model_1 computes the temporal derivatives
% of the seven dependent variables
%
% Parameters shared with other routines
  global ncall
```

```
%
% Model dependent variables
  yla=y(1);
  yua=y(2);
  ysb=y(3);
  ylb=y(4);
  yul=y(5);
  ydl=y(6);
  ymb=y(7);
%
% ODEs
  [c1,r1]=CO2_rate(t);
  dyla=      1/5*(yua-yla)+...
             1/0.75*(ysb-yla)+...
             1/150*(ylb-yla)+...
             1/30*(yul-yla)+...
             c1*exp(r1*(t-1850)));
  dyua=      1/3*(yla-yua);
  dysb=  1/0.75*(yla-ysb);
  dylb=   1/150*(yla-ylb);
  dyul=      1/80*(yla-yul)+...
             1/200*(ydl-yul)+...
             1/5*(ymb-yul);
  dydl=  1/1500*(yul-ydl);
  dymb=    1/10*(yul-ymb);
%
% Derivative vector
  yt(1)=dyla;
  yt(2)=dyua;
  yt(3)=dysb;
  yt(4)=dylb;
  yt(5)=dyul;
  yt(6)=dydl;
  yt(7)=dymb;
  yt=yt';
%
% Increment calls to model_1
  ncall=ncall+1;
%
% End of model_1
  end
```

We can note the following details about Listings 3a and 3b.

- The functions are defined.

R:

```
model_1=function(t,y,parms){
#
# Function model_1 computes the temporal derivatives
# of the seven dependent variables
```

Matlab:

```
function yt=model_1(t,y)
%
% Function model_1 computes the temporal derivatives
% of the seven dependent variables
%
% Parameters shared with other routines
  global ncall
```

The input (RHS) arguments are t, the independent variable t of Eqs. (3.3) to (3.9) and y, the vector of the seven ODE dependent variables of Eqs. (3.3) to (3.9), $y_{ua}, \ldots y_{mb}$.

Also, in Listing 3b, a global area is used to pass the number of executions of model_1.m to and from the calling program of Listing 1b. This special designation is not required in model_1.R of Listing 3a (a basic feature of R). model_1.R does require a third input argument, e.g., parms, to pass parameters to this routine (but here it is unused).

- In order to facilitate the programming by using problem-oriented variable names, the vector of dependent variables y is placed in the scalar variables yla,...,ymb of Eqs. (3.3) to (3.9).

R:

```
#
# Model dependent variables
  yla=y[1];
  yua=y[2];
  ysb=y[3];
  ylb=y[4];
  yul=y[5];
```

```
ydl=y[6];
ymb=y[7];
```

Matlab:

```
%
% Model dependent variables
  yla=y(1);
  yua=y(2);
  ysb=y(3);
  ylb=y(4);
  yul=y(5);
  ydl=y(6);
  ymb=y(7);
```

- Equations (3.3) to (3.10) are programmed.

R:

```
#
# ODEs
  CO2_rate(t,ncase);
  dyla=      1/5*(yua-yla)+
             1/0.75*(ysb-yla)+
             1/150*(ylb-yla)+
             1/30*(yul-yla)+
             c1*exp(r1*(t-1850)));
  dyua=      1/3*(yla-yua);
  dysb=  1/0.75*(yla-ysb);
  dylb=   1/150*(yla-ylb);
  dyul=    1/80*(yla-yul)+
           1/200*(ydl-yul)+
             1/5*(ymb-yul);
  dydl=  1/1500*(yul-ydl);
  dymb=    1/10*(yul-ymb);
```

Matlab:

```
%
% ODEs
  [c1,r1]=CO2_rate(t);
  dyla=      1/5*(yua-yla)+...
             1/0.75*(ysb-yla)+...
             1/150*(ylb-yla)+...
```

```
            1/30*(yul-yla)+...
            c1*exp(r1*(t-1850)));
dyua=       1/3*(yla-yua);
dysb=       1/0.75*(yla-ysb);
dylb=       1/150*(yla-ylb);
dyul=       1/80*(yla-yul)+...
            1/200*(ydl-yul)+...
            1/5*(ymb-yul);
dydl=       1/1500*(yul-ydl);
dymb=       1/10*(yul-ymb);
```

We can note the following details.

(a) $c1$ and $r1$ in Eq. (3.10) are first computed by a call to CO2_rate.R of Listing 4a and CO2_rate.m of Listing 4b (discussed next) using the input argument t. ncase set in Listings 1a and 1b for a specific emissions function is passed as an argument to CO2_rate.R (CO2_rate(t,ncase)) and as a global variable to CO2_rate.m ([c1,r1]=CO2_rate(t) and therefore does not appear in Listing 3b).

(b) Equation (3.4a) is programmed as

R:

```
dyla=       1/5*(yua-yla)+
            1/0.75*(ysb-yla)+
            1/150*(ylb-yla)+
            1/30*(yul-yla)+
            c1*exp(r1*(t-1850)));
```

Matlab:

```
dyla=       1/5*(yua-yla)+...
            1/0.75*(ysb-yla)+...
            1/150*(ylb-yla)+...
            1/30*(yul-yla)+...
            c1*exp(r1*(t-1850)));
```

The lines in R do not have a continuation character(s) since R permits a continuation of any incomplete line, e.g., a line that ends in + is continued to the next line. Matlab has the continuation characters . . . .

The residence time θ_{ua}^{la} is given the numerical value 5 (years). In other words, the mean residence time of CO_2 entering the lower atmosphere from the upper atmosphere is taken as 5. The residence times ($\theta's$) are "hard coded" to make their values explicit (their values are summarized in Table 2).

Note in particular the four terms for the exchange of CO_2 between the lower atmosphere (represented by yla) and the (1) upper atmosphere (yua), (2) short-lived biota (ysb), (3) long-lived biota (ylb) and (4) ocean upper layer (yul). Also, the emissions function of Eq. (3.10) is included in this code for Eq. (3.4a) (c1*exp(r1*(t-1850))).

- The seven temporal derivatives dyla,...,dymb are then placed in a single vector, yt, that is used by integrators ode (Listing 1a) and ode15s (Listing 1b).

R:

```
#
# Derivative vector
  yt=rep(0,7);
  yt[1]=dyla;
  yt[2]=dyua;
  yt[3]=dysb;
  yt[4]=dylb;
  yt[5]=dyul;
  yt[6]=dydl;
  yt[7]=dymb;
```

Note the vector yt is preallocated before it is used (yt=rep(0,7)).

Matlab:

```
%
% Derivative vector
  yt(1)=dyla;
  yt(2)=dyua;
  yt(3)=dysb;
  yt(4)=dylb;
  yt(5)=dyul;
  yt(6)=dydl;
  yt(7)=dymb;
  yt=yt';
```

A declaration of the vector yt is not required (Matlab dynamically defines vectors and arrays when they are first used).

- The counter for the calls to model_1.R and model_1.m is incremented.

R:

```
#
# Increment calls to model_1
  ncall <<- ncall+1;
```

The value of ncall is returned to the main program of Listing 1a with a <<-.

Matlab:

```
%
% Increment calls to model_1
  ncall=ncall+1;
```

The value of ncall is returned to the main program of Listing 1b as a global variable.

- The return of the derivative vector yt is different in R and Matlab.

R:

```
#
# Return derivative vector
  return(list(c(yt)));
#
# End of model_1
  }
```

The derivative vector is returned as a list which is required by ODE integrator ode. c is the vector operator in R. The final } concludes model_1.R (the beginning { is in model_1=function(t,y,parms){.

Matlab:

```
  yt=yt';
```

The derivative vector is returned as a LHS argument, i.e., function yt=model_1(t,y). Note the use of the transpose (yt=yt') since ode15s requires a column derivative vector (rather than a row vector). The final **end** concludes model_1.

This concludes the programming of Eqs. (3.3) to (3.10). The functions that define the CO_2 emissions rate $Q_c(t)$ in Eq. (3.4a) are considered next.

5.4. Emissions Rate Routines

The CO_2 emissions rate is set in the following routines.

Listing 4a: Routine CO2_rate.R for the programming of Eq. (3.10)

R:

```
  CO2_rate=function(t,ncase){
#
# Function CO2_rate returns the constants c1, r1 in the CO2
# source term
#
#   CO2_rate - c1*exp(r1*(t-1850))
#
# for the case ncase.
#
# c1 sets the CO2 ppm at 2013
  c1=4.92e-03;
#
# Base CO2 rate
  r1b=0.01;r1=r1b;r1c=0;
#
# Change the base rate for t > 2010
  if(t>2010){
    if(ncase==1){r1c= 0.0050;}
    if(ncase==2){r1c= 0.0025;}
    if(ncase==3){r1c= 0.0000;}
    if(ncase==4){r1c=-0.0100;}
  }
#
#   Linear interpolation in t between 2010 and 2100
  r1=r1b+r1c*(t-2010)/(2100-2010);
#
# Return c1,r1
```

```
   c1 <<- c1;
   r1 <<- r1;
#
# End of CO2_rate
   }
```

Listing 4b: Routine CO2_rate.m for the programming of Eq. (3.10)

Matlab:

```
   function [c1,r1]=CO2_rate(t)
%
% Function CO2_rate returns the constants c1, r1 in the CO2
% source term
%
%   CO2_rate = c1*exp(r1*(t-1850))
%
% for the case ncase.
%
   global ncase
%
% c1 sets the CO2 ppm at 2013
   c1=4.92e-03;
%
% Base CO2 rate
   r1b=0.01;r1=r1b;r1c=0;
%
% Change the base rate for t > 2010
   if(t>2010)
     if(ncase==1)r1c= 0.0050; end
     if(ncase==2)r1c= 0.0025; end
     if(ncase==3)r1c= 0.0000; end
     if(ncase==4)r1c=-0.0100; end
   end
%
%   Linear interpolation in t between 2010 and 2100
   r1=r1b+r1c*(t-2010)/(2100-2010);
%
% End of CO2_rate
   end
```

We can note the following points about CO2_rate.R and CO2_rate.m.

- The function is defined.

    R:

    ```
    CO2_rate=function(t,ncase){
    #
    # Function CO2_rate returns the constants c1, r1 in the CO2
    # source term
    #
    #   CO2_rate = c1*exp(r1*(t-1850))
    #
    # for the case ncase.
    ```

    ncase is available as the second input argument to select a particular emissions rate. ncase is set numerically in the main program of Listing 1a and is passed to model_1.R of Listing 3a which in turn passes it to CO2_rate.R of Listing 4a.

    Matlab:

    ```
    function [c1,r1]=CO2_rate(t)
    %
    % Function CO2_rate returns the constants c1, r1 in the CO2
    % source term
    %
    %   CO2_rate = c1*exp(r1*(t-1850))
    %
    % for the case ncase.
    %
    global ncase
    ```

    ncase is declared global so that it has the value set in the main program of Listing 1b.
- c_1 is set to a value that gives $c_{la}(2013) = 400$ ppm for the lower atmosphere as measured and reported [12] (this will be observed in the model output discussed subsequently).

R:

```
#
# c1 sets the CO2 ppm at 2013
  c1=4.92e-03;
```

Matlab:

```
%
% c1 sets the CO2 ppm at 2013
  c1=4.92e-03;
```

- r_1 is set to a base value **r1b=0.01** (a %1 annual increase in CO_2 emissions rate in the lower atmosphere).

R:

```
#
# Base CO2 rate
  r1b=0.01;r1=r1b;r1c=0;
```

Matlab:

```
%
% Base CO2 rate
  r1b=0.01;r1=r1b;r1c=0;
```

- From 1850 to 2010, r_1 remains constant. Then for $t > 2010$, this rate is varied according to four cases, **ncase=1,2,3,4**.

R:

```
#
# Change the base rate for t > 2010
  if(t>2010){
    if(ncase==1){r1c= 0.0050;}
    if(ncase==2){r1c= 0.0025;}
    if(ncase==3){r1c= 0.0000;}
    if(ncase==4){r1c=-0.0100;}
  }
```

Matlab:

```
%
% Change the base rate for t > 2010
  if(t>2010)
    if(ncase==1)r1c= 0.0050; end
    if(ncase==2)r1c= 0.0025; end
    if(ncase==3)r1c= 0.0000; end
    if(ncase==4)r1c=-0.0100; end
  end
```

- The change in r_1 after 2010 is linear in t. In this way, the effects of changes in the anthropogenic emissions rate with t can be investigated.

R:

```
#
#    Linear interpolation in t between 2010 and 2100
     r1=r1b+r1c*(t-2010)/(2100-2010);
#
# Return c1,r1
  c1 <<- c1;
  r1 <<- r1;
#
# End of CO2_rate
  }
```

The final } terminates co2_rate.R.

Matlab:

```
%
%    Linear interpolation in t between 2010 and 2100
     r1=r1b+r1c*(t-2010)/(2100-2010);
%
% End of CO2_rate
  end
```

The final end concludes CO2_rate.m.

For $t > 2010$, the base value r1b=0.01 is changed by r1c over the interval $2010 \le t \le 2100$. Thus, for ncase=1, r_1 in Eq. (3.10) is changed from

```
r1=r1b+r1c*(2010-2010)/(2100-2010)=r1b=0.01
```

at 2010 to

```
r1=r1b+r1c*(2100-2010)/(2100-2010)=r1b+r1c=0.01+0.005=0.015
```

at 2100 (that is, a 50% increase in the CO_2 emission rate r1). This accelerating rate of CO_2 emissions is reflected in the actual increases in the rate [4], that is, not only is the CO_2 emissions rate increasing exponentially as reflected in Eq. (3.10), but the rate r_1 is increasing with t.

For ncase=2 the increase in the emissions rate gives

```
r1=r1b+r1c*(2100-2010)/(2100-2010)=r1b+r1c=0.01+0.0025=0.0125
```

at 2100 representing a 50% decrease of the rate for ncase=1. For ncase=3 there is no increase in the emissions rate and

```
r1=r1b+r1c*(2100-2010)/(2100-2010)=r1b+r1c=0.01+0.0000=0.01
```

at $t = 2100$. For ncase=4 there is a decrease in the emissions rate and

```
r1=r1b+r1c*(2100-2010)/(2100-2010)=r1b+r1c=0.01-0.0100=0.00
```

at $t = 2100$. Thus, for ncase=4, the change in the CO_2 emissions rate with t, as reflected in r_1 of Eq. (3.10), is postulated to go to zero by 2100 so that the emissions rate from Eq. (3.10) becomes constant at the value $Q_c(2100) = c_1 e^0$ (possibly the result of a switch in global use of fossil fuels to some other energy technologies). Note, however, that this does not mean the CO_2 emissions rate, $Q_c(t)$ of Eq. (3.10), has decreased to zero; rather, just the exponent r_1 in Eq. (3.10) is postulated to go to zero so that there is no change in $Q_c(t)$ with t at $t = 2100$.

These four cases demonstrate how the model can be used to investigate various scenarios for changes in the emissions rate. Of course, any particular scenario cannot be considered as a prediction (only a projection) of what will actually happen to CO_2 emissions, but the scenarios could be useful for evaluating changes in CO_2 policies and regulations, changes in technology, etc.

In Listings 1a and 1b, ncase is set to a single value. These main programs can easily be modified to cycle through ncase=1,2,3,4. This extension is discussed subsequently.

At this point, we still require a discussion of the routines `splines.R` and `splines.m` (called in the main programs of Listings 1a and 1b) which implement the ocean chemistry. However, we defer the discussion of these routines in order to present some of the output from the main program of Listings 1a and 1b. Once this is covered, we return to the ocean chemistry and the output for the four cases corresponding to `ncase =1,2,3,4`.

5.5. Base Case Model Output

The numerical and graphical output from `model_1_main.R`, `plot_output.R`, `model_1.R` and `CO2_rate.m` of Listings 1a, 2a, 3a and 4a, and `model_1_main.m`, `plot_output.m`, `model_1.m` and `CO2_rate.m` of Listings 1b, 2b, 3b and 4b (for `ncase=1`) are discussed next, beginning with the tabulated numerical output.

Table 3a: Numerical output from the main program of Listing 1a

R:

ncase =   1

t	yla	yul	ppm	Tc	eps	pH	r1
1850	0.0000	0.0000	280.0	1.745	8.800	8.222	0.0100
1860	0.0200	0.0008	285.6	1.749	8.880	8.215	0.0100
1870	0.0378	0.0024	290.6	1.753	8.928	8.209	0.0100
1880	0.0555	0.0045	295.5	1.756	8.983	8.203	0.0100
1890	0.0734	0.0073	300.6	1.759	9.042	8.197	0.0100
1900	0.0919	0.0106	305.7	1.763	9.106	8.191	0.0100
1910	0.1112	0.0145	311.1	1.766	9.175	8.185	0.0100
1920	0.1316	0.0189	316.9	1.770	9.250	8.178	0.0100
1930	0.1535	0.0240	323.0	1.773	9.331	8.172	0.0100
1940	0.1769	0.0298	329.5	1.777	9.419	8.164	0.0100
1950	0.2024	0.0362	336.7	1.781	9.516	8.157	0.0100
1960	0.2300	0.0434	344.4	1.785	9.621	8.149	0.0100
1970	0.2602	0.0514	352.9	1.790	9.733	8.140	0.0100
1980	0.2932	0.0602	362.1	1.795	9.854	8.131	0.0100
1990	0.3295	0.0700	372.3	1.800	9.979	8.121	0.0100
2000	0.3693	0.0809	383.4	1.805	10.106	8.111	0.0100
2010	0.4131	0.0929	395.7	1.811	10.228	8.100	0.0100
2020	0.4667	0.1063	410.7	1.817	10.360	8.086	0.0106

2030	0.5366	0.1218	430.2	1.825	10.533	8.069	0.0111
2040	0.6271	0.1401	455.6	1.835	10.771	8.049	0.0117
2050	0.7446	0.1621	488.5	1.847	11.093	8.023	0.0122
2060	0.8976	0.1890	531.3	1.861	11.516	7.992	0.0128
2070	1.0980	0.2224	587.4	1.878	12.063	7.955	0.0133
2080	1.3624	0.2647	661.5	1.896	12.706	7.910	0.0139
2090	1.7143	0.3187	760.0	1.917	13.489	7.857	0.0144
2100	2.1869	0.3886	892.3	1.940	14.422	7.795	0.0150

```
Atmosphere ppm(2013) =   400.2

GtC Lower Atmosphere(2013) =   725.3

GtC Upper Atmosphere(2013) =   126.7

GtC Long-lived(2013)       = 2498.9

GtC Short-lived(2013)      =  156.8

GtC Upper Layer(2013)      =  987.2

GtC Deep Layer(2013)       = 37227.5

GtC Marine Biota(2013)     =    3.3

ncall =   363
```

Table 3b: Numerical output from the main program of Listing 1b

Matlab:

```
ncase =    1

   t     yla     yul     ppm     Tc     eps     pH     r1
1850  0.0000  0.0000   280.0  1.745  8.800  8.222  0.0100
1860  0.0200  0.0008   285.6  1.749  8.898  8.215  0.0100
1870  0.0378  0.0024   290.6  1.753  8.946  8.209  0.0100
1880  0.0555  0.0045   295.5  1.756  9.000  8.203  0.0100
1890  0.0734  0.0073   300.6  1.759  9.058  8.197  0.0100
1900  0.0919  0.0106   305.7  1.763  9.119  8.191  0.0100
1910  0.1112  0.0145   311.1  1.766  9.186  8.185  0.0100
1920  0.1316  0.0189   316.9  1.769  9.258  8.178  0.0100
```

```
1930  0.1535  0.0240  323.0  1.773   9.337  8.172  0.0100
1940  0.1769  0.0298  329.5  1.777   9.423  8.164  0.0100
1950  0.2024  0.0362  336.7  1.781   9.517  8.157  0.0100
1960  0.2300  0.0434  344.4  1.785   9.620  8.149  0.0100
1970  0.2602  0.0514  352.9  1.790   9.730  8.140  0.0100
1980  0.2932  0.0602  362.1  1.795   9.848  8.131  0.0100
1990  0.3295  0.0700  372.3  1.800   9.972  8.121  0.0100
2000  0.3693  0.0809  383.4  1.805  10.098  8.111  0.0100
2010  0.4131  0.0929  395.7  1.811  10.220  8.100  0.0100
2020  0.4667  0.1063  410.7  1.817  10.354  8.086  0.0106
2030  0.5366  0.1218  430.2  1.825  10.530  8.069  0.0111
2040  0.6271  0.1401  455.6  1.835  10.771  8.049  0.0117
2050  0.7446  0.1621  488.5  1.847  11.096  8.023  0.0122
2060  0.8976  0.1890  531.3  1.861  11.518  7.992  0.0128
2070  1.0980  0.2224  587.4  1.878  12.063  7.955  0.0133
2080  1.3624  0.2647  661.5  1.896  12.706  7.910  0.0139
2090  1.7143  0.3187  760.0  1.917  13.490  7.857  0.0144
2100  2.1869  0.3886  892.3  1.940  14.422  7.795  0.0150


Atmosphere ppm(2013) =   400.2

GtC Lower Atmosphere(2013) =   725.3

GtC Upper Atmosphere(2013) =   126.7

GtC Long-lived(2013)        = 2498.9

GtC Short-lived(2013)       =  156.8

GtC Upper Layer(2013)       =  987.2

GtC Deep Layer(2013)        = 37227.5

GtC Marine Biota(2013)      =    3.3

ncall =  153
```

We can note the following points about this output.

- t covers the interval $1850 \leq t \leq 2100$, as specified in Listings 1a and 1b. Within this interval, the R and Matlab solutions are in close agreement, e.g.,

t	yla	yul	ppm	Tc	eps	pH	r1
(R)							
1850	0.0000	0.0000	280.0	1.745	8.800	8.222	0.0100
(Matlab)							
1850	0.0000	0.0000	280.0	1.745	8.800	8.222	0.0100
(R)							
1900	0.0919	0.0106	305.7	1.763	9.106	8.191	0.0100
(Matlab)							
1900	0.0919	0.0106	305.7	1.763	9.119	8.191	0.0100
(R)							
2000	0.3693	0.0809	383.4	1.805	10.106	8.111	0.0100
(Matlab)							
2000	0.3693	0.0809	383.4	1.805	10.098	8.111	0.0100
(R)							
2100	2.1869	0.3886	892.3	1.940	14.422	7.795	0.0150
(Matlab)							
2100	2.1869	0.3886	892.3	1.940	14.422	7.795	0.0150

- The fractional increase of yla corresponds to $c_{la}(2100) = 280(1 + 2.187) = 892.3$ ppm as reflected in the output at $t = 2100$. Also, the value of $c_{la}(2013) = 400.2$ ppm interpolated from values between 2010 and 2020 corresponds to the observed level of atmospheric CO_2 as discussed previously (when $r_1 = 4.92 \times 10^{-3}$ was programmed in CO2_rate.R and CO2_rate.m of Listings 4a and 4b).

- Additionally, the rate of increase of $c_{la}(t)$ in ppm between 2010 and 2020 is $410.7 - 395.7 = 15$ which is in good agreement with the reported value of 15 ppm CO_2/decade. This rate is dependent on the value of r_1 and increases substantially beyond 2020 (as r_1 in CO2_rate.R,CO2_rate.m of Listings 4a and 4b increases).

- In other words, r_1 is constant at 0.01 until 2010, then increases linearly to $0.01 + 0.005 = 0.0150$ at $t = 2100$ as programmed in CO2_rate.R, CO2_rate.m (Listings 4a and 4b) for ncase=1 and as listed in the last column of Tables 3a and 3b.

- The final value $c_{la}(2100) = 892.3$ ppm is in line with various projections of atmospheric CO_2 levels at $t = 2100$. Of course, this figure results from the increase in r_1 of Eq. (3.10) to 0.015 at $t = 2100$ as discussed previously (in CO2_rate.R, CO2_rate.m, Listings 4a and 4b).

The additional output in Tables 3a and 3b can be visualized with the three plots that follow produced by model_1_main.R and model_1_main.m of Listings 1a and 1b (according to the three selected plots discussed previously). The increase in CO_2 from 280.0 to 892.3 ppm reflects the values in Tables 3a and 3b. The decrease in pH from 8.222 to 7.795 reflects the values in Tables 3a and 3b.

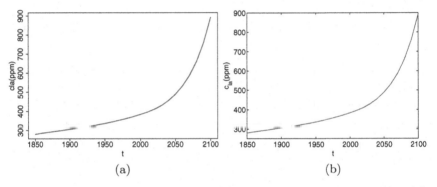

(a)   (b)

Figure 2: (a) $c_{la}(t)$, $1850 \le t \le 2100$ (ppm) from Listing 2a (R). (b) $c_{la}(t)$, $1850 \le t \le 2100$ (ppm) from Listing 2b (Matlab)

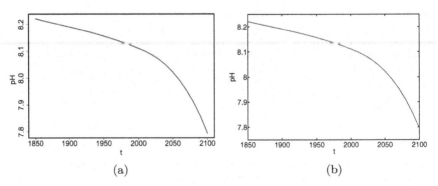

(a)   (b)

Figure 3: (a) $pH(t)$, $1850 \le t \le 2100$ from Listing 2a (R). (b) $pH(t)$, $1850 \le t \le 2100$ from Listing 2b (Matlab)

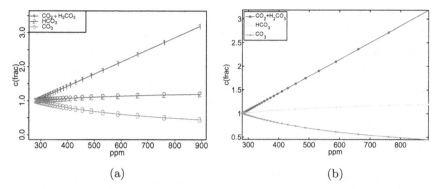

(a)   (b)

Figure 4: (a) Fractional upper layer $CO_2 + H_2CO_3$, HCO_3, CO_3 vs cla (ppm) from Listing 2a (R). (b) Fractional upper layer $CO_2 + H_2CO_3$, HCO_3, CO_3 vs cla (ppm) from Listing 2b (Matlab)

We can note the following points about the pH response:

- The value $pH(t = 2010) = 8.100$ from Tables 3a and 3b is in agreement with the current observed value of 8.1.
- The change in pH from 1850 to 2100, $8.222 - 7.795 = 0.427$, may seem rather small. However, as Eq. (1.1) indicates, pH is a logarithmic function that defines $[H^+]$. Thus, the increase in $[H^+]$ is $10^{0.427} = 2.673$ and logically we could be concerned about the possible detrimental effects this increase in hydrogen ion concentration (a factor of 2.673 reflecting acidification) might have on coral and the ocean biological systems.

The plots of the carbon in the ocean upper layer (Figs. 4a and 4b) indicate that the dissolved CO_2 plus carbonic acid (H_2CO_3) increases substantially with increasing higher atmospheric CO_2 (on a fractional basis relative to the $t = 1850$ values), but the increase in bicarbonate (HCO_3^-) is modest and carbonate (CO_3^{2-}) actually decreases. This rather complex response of the ocean upper layer uptake of CO_2 is an illustration of a counterintuitive response of the model. In this case, since the carbon as $CO_{2(aq)} + H_2CO_3$ in absolute terms is much smaller than the HCO_3^- and CO_3^{2-} (from the calculation of these chemical species in Listings 1a and 1b and discussed next), Figs. 4a and 4b indicate the limited capacity of the ocean upper layer to take up CO_2 with increasing CO_2 in the lower atmosphere. That is, the oceans are not reservoirs of virtually unlimited capacity as often perceived.

This effect of limited ocean uptake of CO_2 can be confirmed by adding the following code for *ppm*, $CO_{2(aq)} + H_2CO_3$, HCO_3^- and CO_3^{2-} to the end of Listings 1a and 1b:

R:

```
#
# Display additional selected output
   cat(sprintf("\n         ppm     co2+h2co3      hco3
          co3\n"));
   for(it in 1:nout){
   cat(sprintf("%10.1f%12.4f%10.4f%10.4f\n",
           ppm[it],co2[it],hco3[it],co3[it]));
   }
```

Matlab:

```
%
% Display additional selected output
   fprintf('\n         ppm     co2+h2co3      hco3          co3\n');
   for(it=1:nout)
     fprintf('%10.1f%12.4f%10.4f%10.4f\n',...
           ppm(it),co2(it),hco3(it),co3(it));
   end
```

A for is used to display numerically the previously calculated *ppm*, $CO_{2(aq)} + H_2CO_3$, HCO_3^- and CO_3^{2-} as a function of t (through the index it). The output from this code is listed next.

Table 4a: Additional numerical output from the main program of Listing 1a

R:

ppm	co2+h2co3	hco3	co3
280.0	0.0094	1.5209	0.2153
285.6	0.0096	1.5271	0.2128
290.6	0.0097	1.5325	0.2106
295.5	0.0099	1.5377	0.2085
300.6	0.0101	1.5428	0.2064

305.7	0.0102	1.5480	0.2043
311.1	0.0104	1.5534	0.2021
316.9	0.0106	1.5589	0.1999
323.0	0.0108	1.5646	0.1975
329.5	0.0110	1.5707	0.1951
336.7	0.0113	1.5770	0.1925
344.4	0.0115	1.5837	0.1898
352.9	0.0118	1.5908	0.1869
362.1	0.0121	1.5984	0.1839
372.3	0.0125	1.6064	0.1806
383.4	0.0128	1.6148	0.1772
395.7	0.0132	1.6237	0.1736
410.7	0.0137	1.6341	0.1694
430.2	0.0144	1.6468	0.1643
455.6	0.0152	1.6622	0.1580
488.5	0.0163	1.6804	0.1506
531.3	0.0178	1.7016	0.1420
587.4	0.0197	1.7259	0.1321
661.5	0.0221	1.7531	0.1211
760.0	0.0254	1.7829	0.1090
892.3	0.0299	1.8144	0.0961

Table 4b: Additional numerical output from the main program of Listing 1b

Matlab:

ppm	co2+h2co3	hco3	co3
280.0	0.0094	1.5209	0.2153
285.6	0.0096	1.5271	0.2128
290.6	0.0097	1.5325	0.2106
295.5	0.0099	1.5377	0.2085
300.6	0.0101	1.5428	0.2064
305.7	0.0102	1.5480	0.2043
311.1	0.0104	1.5534	0.2021
316.9	0.0106	1.5589	0.1999
323.0	0.0108	1.5646	0.1975
329.5	0.0110	1.5707	0.1951
336.7	0.0113	1.5770	0.1925
344.4	0.0115	1.5837	0.1898
352.9	0.0118	1.5908	0.1869
362.1	0.0121	1.5984	0.1839

372.3	0.0125	1.6064	0.1806
383.4	0.0128	1.6148	0.1772
395.7	0.0132	1.6237	0.1736
410.7	0.0137	1.6341	0.1694
430.2	0.0144	1.6468	0.1643
455.6	0.0152	1.6622	0.1580
488.5	0.0163	1.6804	0.1506
531.3	0.0178	1.7016	0.1420
587.4	0.0197	1.7259	0.1321
661.5	0.0221	1.7531	0.1211
760.0	0.0254	1.7829	0.1090
892.3	0.0299	1.8144	0.0961

We can note the following details about this output.

- The numerical values are nearly identical for R and Matlab.
- The dissolved $CO_{2(aq)} + H_2CO_3$ is relatively small indicating that little carbonic acid exists in the ocean upper layer.
- Carbonic acid mainly dissociates into H^+ and HCO_3^- which accounts for most of the H^+ that produces acidification (lower *pH* as indicated in Figs. 3a and 3b).
- The carbonate concentration CO_3^{2-} is relatively small and actually decreases with increasing atmospheric *ppm*.
- The wide variation in concentrations is the reason for using relative concentrations based on normalization by the 1850 values in Figs. 4a and 4b.

As stated previously, this numerical output can be interpreted as the limited capacity of the oceans to take up CO_2 with increasing acidification.

Since global warming and climate change are still controversial and uncertain areas, the primary focus of the model is carbon buildup in the atmosphere and oceans which is being measured quantitatively and is therefore undisputed. However, as explained previously, a temperature sensitivity, co2s, is included in the model to provide an estimate of global warming in the next plot. This sensitivity is defined numerically in the main programs of Listings 1a, 1b as $0.005°C/ppm$ which is then used in plot_out.R, plot_out.m (Listings 2a, 2b), with plot_out[10]=1,plot_out(10)=1. The resulting plots follow.

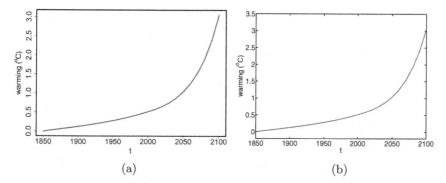

(a)   (b)

Figure 5: (a) Temperature rise (°C), $1850 \leq t \leq 2100$ from Listing 2a (R) and (b) Temperature rise (°C), $1850 \leq t \leq 2100$ from Listing 2b (Matlab)

We can note the following details about Figs. 5a and 5b.

- At $t = 2013$, the temperature rise from $t = 1850$ is approximately $0.6°C = 1.08°F$ which is in line with reported values.
- Most plans that have been put forward for limiting CO_2 emissions have specified a maximum long-term temperature rise of 2°C. From Figs. 5a and 5b, this maximum will occur at about 2080.
- Beyond this point (2080), the temperature rise will be greater than 2°C, e.g., approximately 3°C at 2100 based on `ncase=1` in `model_1.R, model_1.m` (Listings 3a and 3b). Even reducing the emissions rate by 50% in `ncase=2` will cause the temperature rise to exceed the stated goal of 2°C, that is, $ppm(2100) \approx 2.1$, and the temperature rise will continue after 2100 (because of the increasing emissions rate of Eq. (3.4a)).

The preceding outputs in Tables 3a and 3b and Figs. 2a, 2b, 5a and 5b are dependent on the particular values of the mean residence times (the $\theta's$ in Table 2) and in `model_1.R, model_1.m` of Listings 3a and 3b. While the values of the $\theta's$ in Table 2 have substantial uncertainty (they were estimated, either intuitively or from reported reservoir inventories and fluxes [2], [24]), our experience has indicated that the principal features of the model output are not especially sensitive to the specific values of the mean residence times, at least semi-quantitatively. Also, these residence times are not necessarily symmetric with respect to their source and sink reservoirs. For example, $\theta_{dl}^{ul} = 200$ (mean residence time in the upper layer of CO_2 from the deep layer) while $\theta_{ul}^{dl} = 1500$ (mean residence time in the deep layer of CO_2 from the upper layer).

Since the residence times of Table 2 have a maximum ratio of 1500/0.75, we consider Eqs. (3.3) to (3.9) to be moderately stiff. Therefore, we used the stiff ODE integrators ode (in R) and ode15s (in Matlab) in model_1_main.R, model_1_main.m (Listings 1a and 1b). The outputs of Tables 3a and 3b indicate a value of ncall of 363 (R) and 153 (Matlab) which represent a modest computational effort (experience has indicated that depending on the ODE problem, R can have a value of ncall less than for Matlab).

As illustrations of how these residence times of Table 2 were estimated [2], [24]:

- θ_{la}^{sb} for CO_2 transfer from the lower atmosphere to the short-lived biota is taken as 0.75 year since the short-lived biota are typically agricultural crops that grow and are harvested within one year.
- θ_{la}^{lb} for CO_2 transfer from the lower atmosphere to the long-lived biota is taken as 150 years since the long-lived biota are typically trees that grow and are renewed over time on the order of 150 years.
- $\theta_{ul}^{dl} = 1500$ for CO_2 transfer to the deep layer from the upper layer reflects the consensus that the deep ocean will take up and retain CO_2 for many hundreds of years.

While this approach based on mean residence times is clearly a simplification of the phenomena that determine CO_2 levels, it has the advantage that the actual carbon fluxes between the seven reservoirs as depicted by the arrows in Fig. 1 do not have to be specified in the model and computer code. These fluxes are generally difficult to evaluate quantitatively, and are the subject of intensive research.

Also, these mean residence times can easily be programmed as a function of t in model_1.R, model_1.m (Listings 3a and 3b). For example, if we hypothesize that continuing cutting of the world's forests will add to the CO_2 problem, we could investigate this effect by reducing the mean residence time of the long-lived biota, θ_{la}^{lb}, from its value of 150 years in Table 2 to a lower value as a function of t (this variation could be programmed in much the same way as the variation of r_1 in CO2_rate.R, CO2_rate.m with t in Listings 4a and 4b).

The basic ocean chemistry is now considered as implemented in splines.R, splines.m called in model_1_main.R, model_1_main.m of Listings 1a and 1b.

6 OCEAN CHEMISTRY

The effect of atmospheric CO_2 on ocean chemistry is rather complex [9] and varies with location in the Earth's oceans. Here, we adopt a simplified approach, starting with the assumptions that: (a) Only the ocean upper layer is considered (and not the deep layer), and (b) Perfect mixing approximates the ocean upper layer carbon distribution (so that spatial variations are neglected).

To begin, CO_2 dissolves in H_2O to form carbonic acid, H_2CO_3

$$CO_2 + H_2O \rightleftharpoons H_2CO_3. \tag{r1}$$

Two points concerning notation can be noted.

1. The double arrow $\rightleftharpoons$ denotes a reversible chemical reaction (a reaction that can proceed either forward to produce H_2CO_3 or backward to produce CO_2 and H_2O).
2. The reactions are numbered with a prefix r so the preceding reaction is numbered $r1$.

Also, a common convention is to take $[CO_2]$ as the dissolved CO_2, denoted as $[CO_{2(aq)}]$, plus the carbonic acid $[H_2CO_3]$.

Carbonic acid is weak and in turn dissociates into bicarbonate, HCO_3^- and H^+

$$H_2CO_3 \rightleftharpoons H^+ + HCO_3^-. \tag{r2}$$

Note in particular the charge HCO_3^- resulting from the loss of a hydrogen ion, H^+. Bicarbonate in turn dissociates into carbonate, CO_3^{2-} and H^+

$$HCO_3^- \rightleftharpoons H^+ + CO_3^{2-}, \tag{r3}$$

with the charge 2− resulting from the loss of a second hydrogen ion. Reactions (r2) and (r3) produce hydrogen ions and therefore contribute to acidification.

H_2O also dissociates to produce hydrogen ions

$$H_2O \rightleftharpoons H^+ + HO^-. \tag{r4}$$

Additionally, boron hydroxide in seawater dissociates to produce hydrogen ions

$$B(OH)_3 \rightleftharpoons H^+ + B(OH)_4^-. \tag{r5}$$

Other compounds in the oceans also dissociate to produce hydrogen ions, but we will not consider them.

Reactions (r1) to (r5) are included in the mathematical model, mainly to compute the extent of acidification as reflected in $[H^+]$ and pH. To do this, (r1) to (r5) are assumed to be at equilibrium. That is, the rates of the forward and reverse reactions noted by $\rightleftharpoons$ are equal and we can express this condition through the use of equilibrium constants. The assumption of equilibrium chemistry is justified because the time scale of the chemical reactions is much shorter than the time scale of the CO_2 transfer between the reservoirs and thus the accumulation within the reservoirs.

For the following analysis, the equilibrium constants of Bacastow and Keeling [2] (Appendix D) expressed in the units of millimols/liter are used. The relation between gaseous and aqueous CO_2 is (for average ocean surface water of $T = 292.75$ K and chlorinity $Cl = 19.24$ mg/ml[1]).

$$K_0 = \frac{[CO_2]}{p_m} = 3.347 \times 10^{-5}, \tag{6.1}$$

---

[1]Ocean chemistry equilibrium constants are a function of temperature and salinity (chlorinity). The values of K_0, K_1, K_2, K_w, K_b used in the model are nominal values based on a representative average ocean temperature and salinity. Examples of the variation of the equilibrium constant K_0 with temperature and the solubility product K_{sp} of $CaCO_3$ with salinity are discussed in the exercises of Appendix E.

where $[CO_2]$ is the sum of the dissolved CO_2 and carbonic acid as noted previously, and p_m is the gas (air) phase CO_2 partial pressure in ppm.

For (r2)

$$K_1 = \frac{[H^+][HCO_3^-]}{[CO_2]} = 9.747 \times 10^{-4}. \tag{6.2}$$

For (r3)

$$K_2 = \frac{[H^+][CO_3^{2-}]}{[HCO_3^-]} = 8.501 \times 10^{-7}. \tag{6.3}$$

For (r4)

$$K_W = [H^+][OH^-] = 6.46 \times 10^{-9}. \tag{6.4}$$

For (r5)

$$K_B = \frac{[H^+][B(OH)_4^-]}{[B(OH)_3]} = 1.881 \times 10^{-6}, \tag{6.5}$$

where

Constant	Meaning
K_0	Equilibrium constant relating atmospheric and dissolved CO_2 concentrations
K_1, K_2	First and second stoichiometric equilibrium constants of carbonic acid
K_w	Stoichiometric equilibrium constant for water
K_b	Stoichiometric equilibrium constant for boric acid

The starting point in computing $[H^+]$ (and thus acidity) is to use the total alkalinity, A, which expresses the electrical neutrality of ocean water (the positively charged ions must be balanced by the negatively charged ions). For the chemical species discussed previously (in (r1) to (r5)), A is defined as

$$A = [HCO_3^-] + 2\,[CO_3^{2-}] + [B(OH)_4^-] + [OH^-] - [H^+]. \quad (6.6)$$

Values of A [2], [24] are assumed constant (do not change with time t).

If all of the chemical species in Eq. (6.6) are eliminated except $[H^+]$ through the use of Eqs. (6.1) to (6.5), we arrive at a fourth-order polynomial in $[H^+]$, $p_4([H^+])$, (the derivation is given in Appendix C)

$$p_4([H^+]) = \sum_{j=0}^{4} D_i [H^+]^j, \quad (6.7)$$

where

Table 5: Coefficients in the fourth-order polynomial of Eq. (6.7)

$D_0 = 2K_0K_1K_2P_mK_B$
$D_1 = K_0K_1P_mK_B + 2K_0K_1K_2P_m + K_WK_B$
$D_2 = K_0K_1P_m + BK_B + K_W - AK_B$
$D_3 = -K_B - A$
$D_4 = -1$

The fourth-order polynomial of Eq. (6.7) is factored numerically in the routines `splines.R`, `splines.m` (Listings 5a and 5b) by Newton's method

$$[H^+]^{i+1} = [H^+]^i - \frac{p_4([H^+]^i)}{dp_4([H^+]^i)/d[H^+]}, \quad (6.8)$$

where i is the index (counter) for the Newton iterations with the starting (assumed) value $[H^+]^1$ and $dp_4([H^+]^i)/d[H^+]$ is the derivative of $p_4([H^+])$ with respect to $[H^+]$ at iteration i

$$dp_4([H^+])/d[H^+] = \sum_{j=1}^{4} j D_j [H^+]^{j-1}. \quad (6.9)$$

Equations (6.1) to (6.9) define the ocean chemistry, and in particular, $[H^+]$ from Eq. (6.8). These equations are programmed in routines `splines.R`, `splines.m` of Listings 5a and 5b which have the following basic components:

- $[H^+]$ is computed for a series of atmospheric CO_2 concentrations (*ppm*) to produce a table of *ppm*, T_c (total carbon), $[H^+]$, *pH*.
- These tabular values are then approximated by splines to produce functions that can be used for interpolation of the tabulated numerical values. Specifically, splines are polynomials that approximate (represent) paired sets of numbers (x_1, y_1), $(x_2, y_2), \ldots, (x_n, y_n)$, that is $y = f(x)$. Typically, f is a third-order polynomial with the coefficients selected to return exactly (within the computer or machine accuracy or epsilon) the original data (pairs of numbers) and also insure continuity of the first and second derivatives at the data points. Thus, splines provide smooth approximations of the data which is the main reason for their use.
- In summary, tables (of data) are created in `splines.R`, `splines.m`, and the tables are then used to form (spline) functions. This process is implemented with utilities in R and Matlab as illustrated in the following functions (`splines.R`, `splines.m`). The spline functions are then used in the main programs of Listings 1a, 1b for interpolation at a series of times (1850 to 2100) to produce numerical and graphical output (Tables 3a and 3b, Figs. 2a, 2b, 5a and 5b.

Listing 5a: Function `spline.R` for Eqs. (6.1) to (6.9)

R:

```
splines=function(ppm0,dppm,max){
#
# Declare (preallocate) arrays
   ppm=rep(0,max);Tc=rep(0,max);
   hion=rep(0,max);pH=rep(0,max);
#
# First computation (of ppm, Tc, hion, pH)
   k=1;
```

```
#
# Alkalinity, Sarmiento and Gruber, Fig. 8.3.3c, p330;
# Fig. 8.3.6, p334
  A=2.050;
#
# Initial parameters
#
#   ppm
    ppm[1]=ppm0;
#
#   hi (in millimoles/lt units)
    xhi=5.4e-6;
#
# while loop in k
  while(k<=max){
#
# while loop in hi
  dif=1;
  niter=25;
  iter=0;
#
# Polynomial coefficients
  c0=2.0*k0*k1*k2*kb*ppm[k];
  c1=k0*k1*kb*ppm[k]+2.0*k0*k1*k2*ppm[k]+kw*kb;
  c2=k0*k1*ppm[k]+B*kb+kw-A*kb;
  c3=-kb-A;
  c4=-1;
#
  while(dif>1.0e-12){
#
#   Next Newton iteration; test for convergence failure
    iter=iter+1;
    if(iter>niter){
      cat(sprintf(
        "\n iter = %3d, iter exceeds niter for k = %3d\n",
        iter,k));
      break;
    }
    hi=xhi;
#
#   Polynomial
    f=c0+c1*hi+c2*hi^2+c3*hi^3+c4*hi^4;
```

```
#
#    Derivative of polynomial
     df=c1+2*c2*hi+3*c3*hi^2+4*c4*hi^3;
#
#    Newton's method (for new hi)
     xhi=hi-f/df;
#
#    Change in hi (Newton correction)
     dif=abs(xhi-hi);
#
# End of while loop for Newton iteration
   }
#
# h ion
   hion[k]=xhi;
#
# pH
   pH[k]=3-log10(hion[k]);
#
# Total carbon
   Tc[k]=k0*ppm[k]*(1+k1/hion[k]+k1*k2/hion[k]^2);
#
# Continue while loop in k
   k=k+1;
   if(k>max){break;}
   ppm[k]=ppm[k-1]+dppm;
   }
#
# Return table values ppm,Tc,hion,pH
   ppm  <<- ppm;  Tc <<- Tc;
   hion <<- hion; pH <<- pH;
#
# Return spline functions Tcs,hions,pHs
     Tcs=splinefun(ppm,Tc);
   hions=splinefun(ppm,hion);
     pHs=splinefun(ppm,pH);
   Tcs <<- Tcs; hions <<- hions; pHs <<- pHs;
#
# End of splines
   }
```

Listing 5b: Function `spline.m` for Eqs. (6.1) to (6.9)

Matlab:

```
function [Tcs,hions,pHs]=splines(ppm0,dppm,max)
%
% Table values ppm,Tc,hion,pH
  global ppm Tc hion pH
%
% Equilibrium constants, B
  global k0 k1 k2 kb kw B
%
% First computation (of ppm, Tc, hion, pH)
  k=1;
%
% Alkalinity, Sarmiento and Gruber, Fig. 8.3.3c, p330;
% Fig. 8.3.6, p334
  A=2.050;
%
% Initial parameters
%
%   ppm
    ppm(1)=ppm0;
%
%   hi (in millimoles/lt units)
    xhi=5.4e-6;
%
% while loop in k
  while(k<=max)
%
% while loop in hi
  dif=1;
  niter=25;
  iter=0;
%
% Polynomial coefficients
  c0=2.0*k0*k1*k2*kb*ppm(k);
  c1=k0*k1*kb*ppm(k)+2.0*k0*k1*k2*ppm(k)+kw*kb;
  c2=k0*k1*ppm(k)+B*kb+kw-A*kb;
  c3=-kb-A;
  c4=-1;
%
  while(dif>1.0e-12)
```

```
%
%    Next Newton iteration; test for convergence failure
     iter=iter+1;
     if(iter>niter)
       fprintf('\n iter = %3d, iter exceeds niter for
       k = %3d\n',iter,k);
       break;
     end
     hi=xhi;
%
%    Polynomial
     f=c0+c1*hi+c2*hi^2+c3*hi^3+c4*hi^4;
%
%    Derivative of polynomial
     df=c1+2*c2*hi+3*c3*hi^2+4*c4*hi^3;
%
%    Newton's method (for new hi)
     xhi=hi-f/df;
%
%    Change in hi (Newton correction)
     dif=abs(xhi-hi);
%
% End of while loop for Newton iteration
     end
%
% h ion
     hion(k)=xhi;
%
% pH
     pH(k)=3-log10(hion(k));
%
% Total carbon
     Tc(k)=k0*ppm(k)*(1+k1/hion(k)+k1*k2/hion(k)^2);
%
% End of while loop in k
     k=k+1;
     if(k>max)break; end
     ppm(k)=ppm(k-1)+dppm;
     end
%
% Return spline functions Tcs, hions, pHs
     Tcs  =spline(ppm,Tc);
```

```
hions=spline(ppm,hion);
pHs  =spline(ppm,pH);
%
% End of splines
end
```

We can note the following points about `spline.R`, `spline.m`.

- The functions are defined.

R:

```
splines=function(ppm0,dppm,max){
#
# Declare (preallocate) arrays
  ppm=rep(0,max);Tc=rep(0,max);
  hion=rep(0,max);pH=rep(0,max);
#
```

Arrays used in subsequent calculations are defined (preallocated) with **rep**. The arguments of `splines.R` are

Input:

ppm0	inital value of **ppm**
dppm	increment value of **ppm**
max	number of entries in returned tables (set to 16 in Listing 1a)

Output:

No output arguments	`splines.R` returns a set of spline tables and functions as discussed below.

Matlab:

```
function [Tcs,hions,pHs]=splines(ppm0,dppm,max)
%
% Table values ppm,Tc,hion,pH
  global ppm Tc hion pH
```

```
%
% Equilibrium constants, B
  global k0 k1 k2 kb kw B
```

Parameters set in the main program of Listing 1b are declared global so they can be accessed in `splines.m`. The spline tables produced in `splines.m` are returned to the main program of Listing 1b as global. The arguments of `splines.m` are

Input:

ppm0   inital value of ppm

dppm   increment value of ppm

max   number of entries in returned tables
                                  (set to 16 in Listing 1b)

Output: Spline functions

Tcs   total carbon $= [CO_2] + [HCO_3^-] +$
                                  $[CO_3^{2-}]$

hions $[H^+]$

pHs   ocean upper layer pH

When `splines.R`, `splines.m` are called by the main programs of Listings 1a and 1b, a set of tables and functions is constructed, each with 16 values corresponding to the values ppm0, ppm0+dppm, ppm0+2dppm, ..., ppm0+15dppm. These tables and functions are then the basis for spline interpolations used to calculate pH and other variables in Listings 1a and 1b before the numerical solution is displayed numerically and plotted (as in Tables 3a and 3b and Figs. 2a, 2b, 5a and 5b).

• The first entry in the tables subsequently produced is specified; then alkalinity A used in Eqs. (6.1) to (6.7) is defined.

R:

```
#
# First computation (of ppm,Tc,hion,pH)
  k=1;
#
# Alkalinity, Sarmiento and Gruber, Fig. 8.3.3c, p330;
# Fig. 8.3.6, p334
  A=2.050;
```

Matlab:

```
%
% First computation (of ppm,Tc,hion,pH)
  k=1;
%
% Alkalinity, Sarmiento and Gruber, Fig. 8.3.3c, p330;
% Fig. 8.3.6, p334
  A=2.050;
```

- Two parameters in the calculation of pH are initialized.

R:

```
#
# Initial parameters
#
#   ppm
    ppm[1]=ppm0;
#
#   hi (in millimoles/lt units)
    xhi=5.4e-6;
```

Matlab:

```
%
% Initial parameters
%
%   ppm
    ppm(1)=ppm0;
%
%   hi (in millimoles/lt units)
    xhi=5.4e-6;
```

Note the initial ppm[1] = ppm(1) = ppm0 = 200 for the first line of the computed tables and the initial value of $[H^+]$ to start the Newton iteration of Eq. (6.8).

- A while is used to step through the 16 values of ppm, from ppm[1], ppm(1) = 200 to ppm[16], ppm(16) = 200 + (16 − 1)(100) = 1700.

R:

```
#
# while loop in k
  while(k<=max){
```

Matlab:

```
%
% while loop in k
  while(k<=max)
```

where max=16 (through a RHS argument of the function).

- A while is initialized to step through the Newton iterations according to Eq. (6.8).

R:

```
#
# while loop in hi
  dif=1;niter=25;iter=0;
```

Matlab:

```
%
% while loop in hi
  dif=1;niter=25;iter=0;
```

niter is the maximum permitted number of Newton iterations, starting with an initialized iteration counter iter. The difference between two successive values of $[H^+]$ is set to a large initial value, dif, then is reduced as the Newton iterations proceed.

- The coefficients of the fourth-order polynomial of Eq. (6.7) are computed as a function of ppm.

R:

```
#
# Polynomial coefficients
  c0=2.0*k0*k1*k2*kb*ppm[k];
  c1=k0*k1*kb*ppm[k]+2.0*k0*k1*k2*ppm[k]+kw*kb;
  c2=k0*k1*ppm[k]+B*kb+kw-A*kb;
  c3=-kb-A;
  c4=-1;
```

The constants k1 to B are defined numerically in the main program of Listing 1a and are available to spline.R without a special designation (a basic feature of R).

Matlab:

```
%
% Polynomial coefficients
  c0=2.0*k0*k1*k2*kb*ppm(k);
  c1=k0*k1*kb*ppm(k)+2.0*k0*k1*k2*ppm(k)+kw*kb;
  c2=k0*k1*ppm(k)+B*kb+kw-A*kb;
  c3=-kb-A;
  c4=-1;
```

The constants k1 to B are defined numerically in the main program of Listing 1b and are available to spline.m with the previous global designation global k0 k1 k2 kb kw B (from Listing 5b).

- The while for the Newton iterations of Eq. (6.8) continues until the difference between two successive values of $[H^+]$ is less than 1.0e-12. Recall that the initial (trial) $[H^+]$ is xhi=5.4e-6 so the error tolerance of 1.0e-12 is a factor of approximately 10^{-6} below the starting value.

R:

```
#
  while(dif>1.0e-12){
#
#   Next Newton iteration; test for convergence failure
    iter=iter+1;
    if(iter>niter){
      cat(sprintf(
        "\n iter = %3d, iter exceeds niter for k = %3d\n",
        iter,k));
```

```
    break;
  }
  hi=xhi;
```

Matlab:

```
%
  while(dif>1.0e-12)
%
%   Next Newton iteration; test for convergence failure
    iter=iter+1;
    if(iter>niter)
      fprintf('\n iter = %3d, iter exceeds niter for
      k = %3d\n',iter,k);
      break;
    end
    hi=xhi;
```

If the Newton iteration does not converge in `niter` = 25 passes, a warning message is displayed and the iteration is terminated by exiting from the `while` (using `break`). Otherwise, $[H^+]$ = `xhi` is updated and the Newton iteration continues.

- The polynomial of Eq. (6.7) is evaluated (`f`) and its derivative in Eq. (6.9) is computed (`df`).

R:

```
#
#   Polynomial
    f=c0+c1*hi+c2*hi^2+c3*hi^3+c4*hi^4;
#
#   Derivative of polynomial
    df=c1+2*c2*hi+3*c3*hi^2+4*c4*hi^3;
```

Matlab:

```
%
%   Polynomial
    f=c0+c1*hi+c2*hi^2+c3*hi^3+c4*hi^4;
%
%   Derivative of polynomial
    df=c1+2*c2*hi+3*c3*hi^2+4*c4*hi^3;
```

- Newton's method of Eq. (6.8) is used to update $[H^+] = $ xhi and the difference in the sucessive values of $[H^+]$ is computed to determine if the Newton iteration has converged.

R:

```
#
#    Newton's method (for new hi)
     xhi=hi-f/df;
#
#    Change in hi (Newton correction)
     dif=abs(xhi-hi);
#
# End of while loop for Newton iteration
  }
```

The while for the Newton iteration is concluded with a }.

Matlab:

```
%
%    Newton's method (for new hi)
     xhi=hi-f/df;
%
%    Change in hi (Newton correction)
     dif=abs(xhi-hi);
%
% End of while loop for Newton iteration
  end
```

The while for the Newton iteration is concluded with an **end**. The rate of convergence of the Newton iteration can be judged by the value of iter after completion of the while loop. Generally, the quadratic convergence of Newton's method produces a final value of xhi in less than five iterations. This is particularly noteworthy since the starting value of xhi in all cases is taken as a constant, xhi=5.4e-6. If larger deviations of the model variables from their initial values produce convergence failures (as flagged by the error message), better starting values of xhi may be required, for example, that are closer to the previous solution from the outer while loop (previous value of k).

- The while in k continues until max = 16 rows of the spline tables are completed. If k < max, ppm is incremented by dppm = 100 and the next row in the spline tables is initiated.

R:

```
#
# Continue while loop in k
  k=k+1;
  if(k>max){break;}
  ppm[k]=ppm[k-1]+dppm;
  }
```

The final } concludes the while in k (so that the spline tables are complete).

Matlab:

```
%
% End of while loop in k
  k=k+1;
  if(k>max)break; end
  ppm(k)=ppm(k-1)+dppm;
  end
```

The final end concludes the while in k (so that the spline tables are complete).

- The spline tables are returned to the main program of Listing 1a with <<- for R.

R:

```
#
# Return table values ppm,Tc,hion,pH
  ppm  <<- ppm;  Tc <<- Tc;
  hion <<- hion; pH <<- pH;
```

Matlab:

Spline tables returned through global area.

- The approximating spline functions Tcs, hions, pHs are produced from the spline tables Tc, hion, pH by the utilities splinefun.R, spline.m.

R:

```
#
# Return spline functions Tcs,hions,pHs
    Tcs=splinefun(ppm,Tc);
    hions=splinefun(ppm,hion);
    pHs=splinefun(ppm,pH);
    Tcs <<- Tcs; hions <<- hions; pHs <<- pHs;
#
# End of splines
    }
```

The approximating spline functions Tcs, hions, pHs are returned to the main program of Listing 1a with <<-. splines.R is concluded with the final }.

Matlab:

```
%
% Return spline functions Tcs,hions,pHs
    Tcs =spline(ppm,Tc);
    hions=spline(ppm,hion);
    pHs =spline(ppm,pH);
%
% End of splines
    end
```

The approximating spline functions hions, Tcs, pHs are returned to the main program of Listing 1b as the LHS (output) arguments of splines.m.

This concludes the discussion of the base case CO_2 model in R and Matlab. To summarize the routines,

	Listings	Function name
	Listing 1a	main program (R)
	Listing 1b	main program (Matlab)
	Listing 2a	plot_output.R
	Listing 2b	plot_output.m

Listings	Function name
Listing 3a	model_1.R
Listing 3b	model_1.m
Listing 4a	CO2_rate.R
Listing 4b	CO2_rate.m
Listing 5a	spline.R
Listing 5b	spline.m

7 MODEL CASE STUDIES

The operational model (Listings 1a to 5b) can now be used to investigate various scenarios and to compare the output with observational data. We illustrate this use with (1) `ncase=1,2,3,4` in the main programs of Listings 1a, 1b, and (2) comparison of the model output with the atmospheric CO_2 data from Mauna Loa, Hawaii [8].

7.1. Emission Rate Multiple Cases

We start with the revised main programs, Listings 1c and 1d (next), for `ncase=1,2,3,4` (see also `co2_rate.R`, `co2_rate.m` of Listings 4a and 4b for the programming of the CO_2 emission rate parameters c_1, r_1 of Eq. (4.10) for these four cases).

Listing 1c: Revised main program in R for `ncase=1,2,3,4`

R:

```
#
# Delete previous workspaces
  rm(list=ls(all=TRUE))
#
# Access ODE integrator
  library("deSolve");
#
# Access files
  setwd("c:/model_R");
  source("model_1.R");
  source("splines.R");
  source("CO2_rate.R");
  source("plot_output.R");
```

```
#
# Declare (preallocate) arrays
  nout=26;
   r1e=rep(0,nout);  ppm=rep(0,nout); eps=rep(0,nout);
   Tc=rep(0,nout); hion=rep(0,nout);  pH=rep(0,nout);
   co2=rep(0,nout); hco3=rep(0,nout); co3=rep(0,nout);
  boh4=rep(0,nout);
#
# Dimensional ODE dependent variables
  cla=rep(0,nout);cua=rep(0,nout);csb=rep(0,nout);
  clb=rep(0,nout);cul=rep(0,nout);cdl=rep(0,nout);
  cmb=rep(0,nout);
#
# Array for composite plot
  ppm_case=matrix(0,nrow=nout,ncol=4);
#
# Cases 1 to 4 (with composite plot, using abs_out[12]=1;
# below)
  for(ncase in 1:4){
#
# Equilibrium constants (in millimols/lt units)
  k0=0.03347e-03;
  k1=9.747e-04;
  k2=8.501e-07;
  kb=1.881e-06;
  kw=6.46e-09;
#
# Boron
  B=0.409;
#
# Initial conditions
  n=7;
  y0=rep(0,n);
#
# Independent variable for ODE integration
  t0=1850;tf=2100;
  tout=seq(from=t0,to=tf,by=(tf-t0)/(nout-1));
  ncall=0;
#
# Initialized carbon reservoirs (GtC)
  c0la=0.85*597;
  c0ua=0.15*597;
```

```
c0sb=110;
c0lb=2190;
c0ul=900;
c0dl=37100;
c0mb=3;
#
# ODE integration
out=ode(func=model_1,times=tout,y=y0);
#
# Set up spline interpolation
ppm0=200;dppm=100;max=16;
splines(ppm0,dppm,max);
#
# Display selected output
cat(sprintf("\n ncase = %4d\n",ncase));
cat(sprintf(
"\n    t      yla     yul     ppm     Tc      eps     pH
r1\n"));
for(it in 1:nout){
#
# CO2 emissions rate (in lower atmosphere)
CO2_rate(tout[it],ncase);
r1e[it]=r1;
#
# CO2 ppm (in lower atmosphere)
ppm[it]=280*(1+out[it,2]);
#
# Total carbon
Tc[it]=Tcs(ppm[it]);
#
# Evasion factor
if(it==1){eps[1]=8.8;}
if(it >1){
    num=(ppm[it]-ppm[it-1])/ppm[it-1];
    den=( Tc[it]- Tc[it-1])/ Tc[it-1];
    eps[it]=num/den;
}
#
# Hydrogen ion
hion[it]=hions(ppm[it]);
#
# pH
```

```
  pH[it]=pHs(ppm[it]);
#
# CO2 (CO2 + H2CO3 in upper layer)
  co2[it]=k0*ppm[it];
#
# Bicarbonate
  hco3[it]=k0*k1*ppm[it]/hion[it];
#
# Carbonate
  co3[it]=k0*k1*k2*ppm[it]/hion[it]^2;
#
# Boron
  boh4[it]=B/(1+hion[it]/kb);
#
# Selected output
  cat(sprintf("%5.0f%8.4f%8.4f%8.1f%8.3f%8.3f%8.3f%8.4f\n",
    out[it,1],out[it,2],out[it,6],ppm[it],Tc[it],eps[it],
    pH[it],r1e[it]));
#
# CO2 GtC (in all reservoirs) vs t
  cla[it]=c0la*(1+out[it,2]);
  cua[it]=c0ua*(1+out[it,3]);
  csb[it]=c0sb*(1+out[it,4]);
  clb[it]=c0lb*(1+out[it,5]);
  cul[it]=c0ul*(1+out[it,6]);
  cdl[it]=c0dl*(1+out[it,7]);
  cmb[it]=c0mb*(1+out[it,8]);
#
# Multiple cases
  ppm_case[it,ncase]=ppm[it];
  }
#
# CO2 ppm at 2013 (by linear interpolation between 2010
# and 2020)
  p2013=ppm[17]+(ppm[18]-ppm[17])*(2013-2010)/(2020-2010);
#
# GtC (in all reservoirs) at 2013
  cla2013=cla[17]+(cla[18]-cla[17])*(2013-2010)/(2020-2010);
  cua2013=cua[17]+(cua[18]-cua[17])*(2013-2010)/(2020-2010);
  clb2013=clb[17]+(clb[18]-clb[17])*(2013-2010)/(2020-2010);
  csb2013=csb[17]+(csb[18]-csb[17])*(2013-2010)/(2020-2010);
  cul2013=cul[17]+(cul[18]-cul[17])*(2013-2010)/(2020-2010);
```

```
   cdl2013=cdl[17]+(cdl[18]-cdl[17])*(2013-2010)/(2020-2010);
   cmb2013=cmb[17]+(cmb[18]-cmb[17])*(2013-2010)/(2020-2010);
#
# GtC at 2013 output
   cat(sprintf("\n\n Atmosphere ppm(2013) = %6.1f\n",p2013));
   cat(sprintf("\n GtC Lower Atmosphere(2013) = %6.1f\n",
     cla2013));
   cat(sprintf("\n GtC Upper Atmosphere(2013) = %6.1f\n",
     cua2013));
   cat(sprintf("\n GtC Long-Lived(2013)       = %6.1f\n",
     clb2013));
   cat(sprintf("\n GtC Short-Lived(2013)      = %6.1f\n",
     csb2013));
   cat(sprintf("\n GtC Upper Layer(2013)      = %6.1f\n",
     cul2013));
   cat(sprintf("\n GtC Deep Layer(2013)       = %6.1f\n",
     cdl2013));
   cat(sprintf("\n GtC Marine Biota(2013)     = %6.1f\n",
     cmb2013));
   cat(sprintf("\n ncall = %4d\n",ncall));
#
# CO2 sensitivity (C^o/ppm CO2)
   co2s=0.005;
#
# End for in ncase
   }
#
# Select plots (0 - no plot, 1 - selected plot)
#
#   Default is no plots
   plot_out=rep(0,20);
#
#   Summary of available plots in plot_output.m
#
#   1 - c_{la}(ppm) vs t
#
#   2 - eps vs t
#
#   3 - pH vs t
#
#   4 - CO_2 + H_2CO_3 (millimols/liter) vs ppm
#
```

```
#    5 - HCO_3^- bicarbonate (millimols/liter) vs ppm
#
#    6 - CO_3^{2-} carbonate (millimols/liter) vs ppm
#
#    7 - B(OH)^-_4 (millimols/liter) vs ppm
#
#    8 - CO_2+H_2CO_3,HCO_3,CO_3 (fraction of 1850 values) vs
#        ppm
#
#    9 - y(1) (lower atmosphere), y(5) (ocean upper layer),
#        y(6) (ocean deep layer) (fractional change from 1850)
#        vs t
#
#   10 - co2s*(c_{la}(t) - c_{la}(1850)) (C^o/ppm*ppm = C^o)
#        vs t (to give an indication of warming)
#
#   11 - mass C vs t, cla, cua, csb, clb, cul, cdl, cmb (GtC)
#        vs t (semilog plot because of differences in magni-
#        tudes)
#
#   12 - composite plot of ppm CO2 in lower atmosphere vs t
#        (use only for four consecutive cases ncase=1,2,3,4)
#
#   Select plots
     plot_out[12]=1;
#
# Plot output
     plot_output(plot_out);
```

Listing 1d: Revised main program in Matlab

Matlab:

```
%
% Clear previous files
  clear all
  clc
%
% General parameters (shared with other routines)
  global ncall ncase
%
% Spline tables
```

```
  global Tc hion pH
%
% Variables for plotting
  global t      y    ppm    co2    hco3    co3   boh4...
         cla    cua    csb    clb    cul    cdl   cmb...
         co2s   eps    ppm_case
%
% Equilibrium constants, B
  global k0 k1 k2 kb kw B
%
% Cases 1 to 4 (with composite plot, using abs_out[12]=1;
% below)
  for(ncase=1:4)
%
% Equilibrium constants (in millimols/lt units)
  k0=0.03347e-03;
  k1=9.747e-04;
  k2=8.501e-07;
  kb=1.881e-06;
  kw=6.46e-09;
%
% Boron
  B=0.409;
%
% Initial conditions
  n=7;
  y0=zeros(1,n);
%
% Independent variable for ODE integration
  t0=1850;tf=2100;nout=26;
  tout=[t0:10:tf]';
  ncall=0;
%
% Initialized carbon reservoirs (GtC)
  c0la=0.85*597;
  c0ua=0.15*597;
  c0sb=110;
  c0lb=2190;
  c0ul=900;
  c0dl=37100;
  c0mb=3;
%
```

```
% ODE integration
  reltol=1.0e-06;abstol=1.0e-06;
  options=odeset('RelTol',reltol,'AbsTol',abstol);
  [t,y]=ode15s(@model_1,tout,y0,options);
%
% Set up spline interpolation
  ppm0=200;dppm=100;max=16;
  [Tcs,hions,pHs]=splines(ppm0,dppm,max);
%
% Display selected output
  fprintf('\n ncase = %4d\n',ncase);
  fprintf(
  '\n    t      yla      yul      ppm      Tc      eps      pH
  r1\n')
  for it=1:nout
%
%    CO2 emissions rate (in lower atmosphere)
     [c1,r1]=CO2_rate(t(it));
     r1_t(it)=r1;
%
%    CO2 ppm (in lower atmosphere)
     ppm(it)=280*(1+y(it,1));
%
% Total carbon
  Tc(it)=ppval(Tcs,ppm(it));
%
% Evasion factor
  if(it==1)eps(1)=8.8;end
  if(it >1)
    num=(ppm(it)-ppm(it-1))/ppm(it-1);
    den=( Tc(it)- Tc(it-1))/ Tc(it-1);
    eps(it)=num/den;
  end
%
%    Hydrogen ion
     hion(it)=ppval(hions,ppm(it));
%
%    pH
     pH(it)=ppval(pHs,ppm(it));
%
%    CO2 (CO2 + H2CO3 in upper layer)
     co2(it)=k0*ppm(it);
```

```
%
%   Bicarbonate
    hco3(it)=k0*k1*ppm(it)/hion(it);
%
%   Carbonate
    co3(it)=k0*k1*k2*ppm(it)/hion(it)^2;
%
%   Boron
    boh4(it)=B/(1+hion(it)/kb);
%
%   Selected output
    fprintf('%5.0f%8.4f%8.4f%8.1f%8.3f%8.3f%8.3f%8.4f\n',...
      t(it),y(it,1),y(it,5),ppm(it),Tc(it),eps(it),pH(it),
      r1_t(it));
%
% CO2 GtC (in all reservoirs) vs t
    cla(it)=c0la*(1+y(it,1));
    cua(it)=c0ua*(1+y(it,2));
    csb(it)=c0sb*(1+y(it,3));
    clb(it)=c0lb*(1+y(it,4));
    cul(it)=c0ul*(1+y(it,5));
    cdl(it)=c0dl*(1+y(it,6));
    cmb(it)=c0mb*(1+y(it,7));
%
% Multiple cases
    ppm_case(it,ncase)=ppm(it);
    end
%
% CO2 ppm at 2013 (by linear interpolation between 2010
% and 2020)
    p2013=ppm(17)+(ppm(18)-ppm(17))*(2013-2010)/(2020-2010);
%
% GtC (in all reservoirs) at 2013
    cla2013=cla(17)+(cla(18)-cla(17))*(2013-2010)/(2020-2010);
    cua2013=cua(17)+(cua(18)-cua(17))*(2013-2010)/(2020-2010);
    clb2013=clb(17)+(clb(18)-clb(17))*(2013-2010)/(2020-2010);
    csb2013=csb(17)+(csb(18)-csb(17))*(2013-2010)/(2020-2010);
    cul2013=cul(17)+(cul(18)-cul(17))*(2013-2010)/(2020-2010);
    cdl2013=cdl(17)+(cdl(18)-cdl(17))*(2013-2010)/(2020-2010);
    cmb2013=cmb(17)+(cmb(18)-cmb(17))*(2013-2010)/(2020-2010);
%
% GtC at 2013 output
```

```
    fprintf('\n\n Atmosphere ppm(2013) = %6.1f\n',p2013);
    fprintf('\n GtC Lower Atmosphere(2013) = %6.1f\n',cla2013);
    fprintf('\n GtC Upper Atmosphere(2013) = %6.1f\n',cua2013);
    fprintf('\n GtC Long-Lived(2013)       = %6.1f\n',clb2013);
    fprintf('\n GtC Short-Lived(2013)      = %6.1f\n',csb2013);
    fprintf('\n GtC Upper Layer(2013)      = %6.1f\n',cul2013);
    fprintf('\n GtC Deep Layer(2013)       = %6.1f\n',cdl2013);
    fprintf('\n GtC Marine Biota(2013)     = %6.1f\n',cmb2013);
    fprintf('\n ncall = %4d\n',ncall);
%
% CO2 sensitivity (C^o/ppm CO2)
    co2s=0.005;
%
% Next case
    end
%
% Select plots (0 - no plot, 1 - selected plot)
%
%    Default is no plots
     plot_out=zeros(20,1);
%
% Summary of available plots in plot_output.m
%
%    1 - c_{la}(ppm) vs t
%
%    2 - eps vs t
%
%    3 - pH vs t
%
%    4 - CO_2 + H_2CO_3 (millimols/liter) vs ppm
%
%    5 - HCO_3 - bicarbonate (millimols/liter) vs ppm
%
%    6 - CO_3^{2-} carbonate (millimols/liter) vs ppm
%
%    7 - B(OH)^-_4 (millimols/liter) vs ppm
%
%    8 - CO_2+H_2CO_3,HCO_3,CO_3 (fraction of 1850 values) vs
%        ppm
%
%    9 - y(1) (lower atmosphere), y(5) (ocean upper layer),
%        y(6) (ocean deep layer) (fractional change from 1850)
```

```
%         vs t
%
%  10 - co2s*(c_{la}(t) - c_{la}(1850)) (C^o/ppm*ppm = C^o)
%         vs t (to give an indication of warming)
%
%  11 - mass C vs t, cla, cua, csb, clb, cul, cdl, cmb (GtC)
%         vs t (semilog plot because of differences in magni-
%         tudes)
%
%  12 - composite plot of ppm CO2 in lower atmosphere vs t
%         (use only for four consecutive cases ncase=1,2,3,4)
   plot_out(12)=1;
%
% Plot output
  plot_output(plot_out);
```

We can note the following details about Listings 1c and 1d, with emphasis on the differences with Listings 1a and 1b.

- For ncase=1,2,3,4 taken together, plot_output[12] or plot_output(12) is used for a composite plot of ppm vs t (see the comments in Listings 2a and 2b for additional details). This case is used with for(ncase in 1:4){ (R in Listing 1c) or for ncase=1:4 (Matlab in Listing 1d).

R:

```
#
# Cases 1 to 4 (with composite plot, using abs_out[12]=1;
# below)
  for(ncase in 1:4){
```

Matlab:

```
%
% Cases 1 to 4 (with composite plot, using abs_out[12]=1;
% below)
  for(ncase=1:4)
```

In Listing 1c

```
#
# Array for composite plot
  ppm_case=matrix(0,nrow=nout,ncol=4);
```

defines a matrix **ppm_case** for the composite plot from **plot_output[12]** (it is passed automatically to **plot_out** of Listing 2a for plotting as a basic feature of R).

In Listing 1d, this array is designated global so that it is passed to **plot_output** of Listing 2b (it does not have to be declared (dimensioned) in Matlab). Also, **ppm_case** is added to the global area of **plot_output** in Listing 2b.

```
%
% Variables for plotting
  global t     y     ppm   co2  hco3   co3   boh4...
         cla   cua   csb   clb  cul    cdl   cmb...
         co2s  eps   ppm_case
```

- Array **ppm_case** contains **ppm** as a function of t (index **it**) and **ncase**.

  R:

```
#
# Multiple cases
  ppm_case[it,ncase]=ppm[it];
```

  Matlab:

```
%
% Multiple cases
  ppm_case(it,ncase)=ppm(it);
```

- The **for** in **ncase** is terminated as } (Listing 1c) and **end** (Listing 1d).

R:

```
#
# CO2 sensitivity (C^o/ppm CO2)
  co2s=0.005;
#
# End for in ncase
  }
```

Matlab:

```
%
% CO2 sensitivity (C^o/ppm CO2)
  co2s=0.005;
%
% End for in ncase
  end
```

- The composite plots are produced with plot_out[12]=1 (Listing 1c) or plot_out(12)=1 (Listing 1d).

R:

```
#
#  12 - composite plot of ppm CO2 in lower atmosphere vs t
#       (use only for four consecutive cases ncase=1,2,3,4)
#
#  Select plots
   plot_out[12]=1;
```

Matlab:

```
%
%  12 - composite plot of ppm CO2 in lower atmosphere vs t
%       (use only for four consecutive cases ncase=1,2,3,4)
   plot_out(12)=1;
```

Other composite plots can be produced using similar steps and adding the plotting to plot_output.R, plot_output.m of Listings 2a and 2b, e.g., plot_out[13]=1, plot_out(13)=1, etc.

The numerical output for ncase=1,2,3,4 is in Tables 5a and 5b. The graphical output is in Figs. 6a and 6b.

Table 6a: Numerical output from the main program of Listing 1c

ncase =   1

t	yla	yul	ppm	Tc	eps	pH	r1
1850	0.0000	0.0000	280.0	1.745	8.800	8.222	0.0100
1860	0.0200	0.0008	285.6	1.749	8.880	8.215	0.0100
1870	0.0378	0.0024	290.6	1.753	8.928	8.209	0.0100
1880	0.0555	0.0045	295.5	1.756	8.983	8.203	0.0100
1890	0.0734	0.0073	300.6	1.759	9.042	8.197	0.0100
1900	0.0919	0.0106	305.7	1.763	9.106	8.191	0.0100
1910	0.1112	0.0145	311.1	1.766	9.175	8.185	0.0100
1920	0.1316	0.0189	316.9	1.770	9.250	8.178	0.0100
1930	0.1535	0.0240	323.0	1.773	9.331	8.172	0.0100
1940	0.1769	0.0298	329.5	1.777	9.419	8.164	0.0100
1950	0.2024	0.0362	336.7	1.781	9.516	8.157	0.0100
1960	0.2300	0.0434	344.4	1.785	9.621	8.149	0.0100
1970	0.2602	0.0514	352.9	1.790	9.733	8.140	0.0100
1980	0.2932	0.0602	362.1	1.795	9.854	8.131	0.0100
1990	0.3295	0.0700	372.3	1.800	9.979	8.121	0.0100
2000	0.3693	0.0809	383.4	1.805	10.106	8.111	0.0100
2010	0.4131	0.0929	395.7	1.811	10.228	8.100	0.0100
2020	0.4667	0.1063	410.7	1.817	10.360	8.086	0.0106
2030	0.5366	0.1218	430.2	1.825	10.533	8.069	0.0111
2040	0.6271	0.1401	455.6	1.835	10.771	8.049	0.0117
2050	0.7446	0.1621	488.5	1.847	11.093	8.023	0.0122
2060	0.8976	0.1890	531.3	1.861	11.516	7.992	0.0128
2070	1.0980	0.2224	587.4	1.878	12.063	7.955	0.0133
2080	1.3624	0.2647	661.5	1.896	12.706	7.910	0.0139
2090	1.7143	0.3187	760.0	1.917	13.489	7.857	0.0144
2100	2.1869	0.3886	892.3	1.940	14.422	7.795	0.0150

Atmosphere ppm(2013) =   400.2

GtC Lower Atmosphere(2013) =   725.3

GtC Upper Atmosphere(2013) =   126.7

GtC Long-lived(2013)   = 2498.9

GtC Short-lived(2013)   =   156.8

GtC Upper Layer(2013)   = 987.2

GtC Deep Layer(2013)   = 37227.5

GtC Marine Biota(2013)   =   3.3

ncall =   363

ncase =   2

t	yla	yul	ppm	Tc	eps	pH	r1
1850	0.0000	0.0000	280.0	1.745	8.800	8.222	0.0100
1860	0.0200	0.0008	285.6	1.749	8.880	8.215	0.0100
1870	0.0378	0.0024	290.6	1.753	8.928	8.209	0.0100
1880	0.0555	0.0045	295.5	1.756	8.983	8.203	0.0100
1890	0.0734	0.0073	300.6	1.759	9.042	8.197	0.0100
1900	0.0919	0.0106	305.7	1.763	9.106	8.191	0.0100
1910	0.1112	0.0145	311.1	1.766	9.175	8.185	0.0100
1920	0.1316	0.0189	316.9	1.770	9.250	8.178	0.0100
1930	0.1535	0.0240	323.0	1.773	9.331	8.172	0.0100
1940	0.1769	0.0298	329.5	1.777	9.419	8.164	0.0100
1950	0.2024	0.0362	336.7	1.781	9.516	8.157	0.0100
1960	0.2300	0.0434	344.4	1.785	9.621	8.149	0.0100
1970	0.2602	0.0514	352.9	1.790	9.733	8.140	0.0100
1980	0.2932	0.0602	362.1	1.795	9.854	8.131	0.0100
1990	0.3295	0.0700	372.3	1.800	9.979	8.121	0.0100
2000	0.3693	0.0809	383.4	1.805	10.106	8.111	0.0100
2010	0.4131	0.0929	395.7	1.811	10.228	8.100	0.0100
2020	0.4640	0.1063	409.9	1.817	10.349	8.087	0.0103
2030	0.5252	0.1213	427.1	1.824	10.491	8.072	0.0106
2040	0.5987	0.1385	447.6	1.832	10.673	8.055	0.0108
2050	0.6870	0.1583	472.4	1.842	10.904	8.035	0.0111
2060	0.7935	0.1812	502.2	1.852	11.190	8.013	0.0114
2070	0.9222	0.2081	538.2	1.864	11.535	7.987	0.0117
2080	1.0784	0.2397	582.0	1.876	11.950	7.958	0.0119
2090	1.2688	0.2772	635.3	1.890	12.404	7.925	0.0122
2100	1.5018	0.3220	700.5	1.905	12.921	7.888	0.0125

Atmosphere ppm(2013) = 400.0

GtC Lower Atmosphere(2013) = 724.8

GtC Upper Atmosphere(2013) =   126.6

GtC Long-lived(2013)   = 2498.9

GtC Short-lived(2013)   =   156.7

GtC Upper Layer(2013)   =   987.2

GtC Deep Layer(2013)   = 37227.5

GtC Marine Biota(2013)   =   3.3

ncall =   362

ncase =   3

t	yla	yul	ppm	Tc	eps	pH	r1
1850	0.0000	0.0000	280.0	1.745	8.800	8.222	0.0100
1860	0.0200	0.0008	285.6	1.749	8.880	8.215	0.0100
1870	0.0378	0.0024	290.6	1.753	8.928	8.209	0.0100
1880	0.0555	0.0045	295.5	1.756	8.983	8.203	0.0100
1890	0.0734	0.0073	300.6	1.759	9.042	8.197	0.0100
1900	0.0919	0.0106	305.7	1.763	9.106	8.191	0.0100
1910	0.1112	0.0145	311.1	1.766	9.175	8.185	0.0100
1920	0.1316	0.0189	316.9	1.770	9.250	8.178	0.0100
1930	0.1535	0.0240	323.0	1.773	9.331	8.172	0.0100
1940	0.1769	0.0298	329.5	1.777	9.419	8.164	0.0100
1950	0.2024	0.0362	336.7	1.781	9.516	8.157	0.0100
1960	0.2300	0.0434	344.4	1.785	9.621	8.149	0.0100
1970	0.2602	0.0514	352.9	1.790	9.733	8.140	0.0100
1980	0.2932	0.0602	362.1	1.795	9.854	8.131	0.0100
1990	0.3295	0.0700	372.3	1.800	9.979	8.121	0.0100
2000	0.3693	0.0809	383.4	1.805	10.106	8.111	0.0100
2010	0.4131	0.0929	395.7	1.811	10.228	8.100	0.0100
2020	0.4614	0.1062	409.2	1.817	10.338	8.088	0.0100
2030	0.5146	0.1208	424.1	1.823	10.452	8.075	0.0100
2040	0.5733	0.1370	440.5	1.830	10.588	8.061	0.0100
2050	0.6380	0.1549	458.6	1.837	10.746	8.046	0.0100
2060	0.7094	0.1747	478.6	1.844	10.927	8.031	0.0100
2070	0.7883	0.1965	500.7	1.852	11.131	8.014	0.0100
2080	0.8754	0.2205	525.1	1.859	11.354	7.996	0.0100

```
2090  0.9715  0.2471   552.0   1.868  11.603   7.978  0.0100
2100  1.0777  0.2765   581.8   1.876  11.872   7.958  0.0100
```

Atmosphere ppm(2013) =   399.7

GtC Lower Atmosphere(2013) =   724.4

GtC Upper Atmosphere(2013) =   126.6

GtC Long-lived(2013)   = 2498.9

GtC Short-lived(2013)   =   156.6

GtC Upper Layer(2013)   =   987.2

GtC Deep Layer(2013)   = 37227.5

GtC Marine Biota(2013)   =   3.3

ncall =   210

ncase =   4

t	yla	yul	ppm	Tc	eps	pH	r1
1850	0.0000	0.0000	280.0	1.745	8.800	8.222	0.0100
1860	0.0200	0.0008	285.6	1.749	8.880	8.215	0.0100
1870	0.0378	0.0024	290.6	1.753	8.928	8.209	0.0100
1880	0.0555	0.0045	295.5	1.756	8.983	8.203	0.0100
1890	0.0734	0.0073	300.6	1.759	9.042	8.197	0.0100
1900	0.0919	0.0106	305.7	1.763	9.106	8.191	0.0100
1910	0.1112	0.0145	311.1	1.766	9.175	8.185	0.0100
1920	0.1316	0.0189	316.9	1.770	9.250	8.178	0.0100
1930	0.1535	0.0240	323.0	1.773	9.331	8.172	0.0100
1940	0.1769	0.0298	329.5	1.777	9.419	8.164	0.0100
1950	0.2024	0.0362	336.7	1.781	9.516	8.157	0.0100
1960	0.2300	0.0434	344.4	1.785	9.621	8.149	0.0100
1970	0.2602	0.0514	352.9	1.790	9.733	8.140	0.0100
1980	0.2932	0.0602	362.1	1.795	9.854	8.131	0.0100
1990	0.3295	0.0700	372.3	1.800	9.979	8.121	0.0100
2000	0.3693	0.0809	383.4	1.805	10.106	8.111	0.0100
2010	0.4131	0.0929	395.7	1.811	10.228	8.100	0.0100

```
2020  0.4518  0.1059  406.5  1.815  10.298  8.090  0.0089
2030  0.4789  0.1192  414.1  1.819  10.324  8.083  0.0078
2040  0.4957  0.1323  418.8  1.821  10.335  8.079  0.0067
2050  0.5032  0.1450  420.9  1.822  10.332  8.077  0.0056
2060  0.5028  0.1569  420.8  1.822  10.315  8.077  0.0044
2070  0.4961  0.1678  418.9  1.821  10.288  8.079  0.0033
2080  0.4843  0.1776  415.6  1.819  10.253  8.082  0.0022
2090  0.4690  0.1863  411.3  1.817  10.213  8.086  0.0011
2100  0.4514  0.1938  406.4  1.815  10.172  8.090  0.0000
```

Atmosphere ppm(2013) =   398.9

GtC Lower Atmosphere(2013) =   723.0

GtC Upper Atmosphere(2013) =   126.5

GtC Long-lived(2013)   = 2498.7

GtC Short-lived(2013)   = 156.4

GtC Upper Layer(2013)   = 987.1

GtC Deep Layer(2013)   = 37227.5

GtC Marine Biota(2013)   =   3.3

ncall =   489

Table 6b: Numerical output from the main program of Listing 1d

ncase =   1

t	yla	yul	ppm	Tc	eps	pH	r1
1850	0.0000	0.0000	280.0	1.745	8.800	8.222	0.0100
1860	0.0200	0.0008	285.6	1.749	8.898	8.215	0.0100
1870	0.0378	0.0024	290.6	1.753	8.946	8.209	0.0100
1880	0.0555	0.0045	295.5	1.756	9.000	8.203	0.0100
1890	0.0734	0.0073	300.6	1.759	9.058	8.197	0.0100
1900	0.0919	0.0106	305.7	1.763	9.119	8.191	0.0100
1910	0.1112	0.0145	311.1	1.766	9.186	8.185	0.0100
1920	0.1316	0.0189	316.9	1.769	9.258	8.178	0.0100
1930	0.1535	0.0240	323.0	1.773	9.337	8.172	0.0100
1940	0.1769	0.0298	329.5	1.777	9.423	8.164	0.0100

1950	0.2024	0.0362	336.7	1.781	9.517	8.157	0.0100
1960	0.2300	0.0434	344.4	1.785	9.620	8.149	0.0100
1970	0.2602	0.0514	352.9	1.790	9.730	8.140	0.0100
1980	0.2932	0.0602	362.1	1.795	9.848	8.131	0.0100
1990	0.3295	0.0700	372.3	1.800	9.972	8.121	0.0100
2000	0.3693	0.0809	383.4	1.805	10.098	8.111	0.0100
2010	0.4131	0.0929	395.7	1.811	10.220	8.100	0.0100
2020	0.4667	0.1063	410.7	1.817	10.354	8.086	0.0106
2030	0.5366	0.1218	430.2	1.825	10.530	8.069	0.0111
2040	0.6271	0.1401	455.6	1.835	10.771	8.049	0.0117
2050	0.7446	0.1621	488.5	1.847	11.096	8.023	0.0122
2060	0.8976	0.1890	531.3	1.861	11.518	7.992	0.0128
2070	1.0980	0.2224	587.4	1.878	12.063	7.955	0.0133
2080	1.3624	0.2647	661.5	1.896	12.706	7.910	0.0139
2090	1.7143	0.3187	760.0	1.917	13.490	7.857	0.0144
2100	2.1869	0.3886	892.3	1.940	14.422	7.795	0.0150

Atmosphere ppm(2013) = 400.2

GtC Lower Atmosphere(2013) = 725.3

GtC Upper Atmosphere(2013) = 126.7

GtC Long-lived(2013)   = 2498.9

GtC Short-lived(2013)   = 156.8

GtC Upper Layer(2013)   = 987.2

GtC Deep Layer(2013)   = 37227.5

GtC Marine Biota(2013)   =   3.3

ncall = 153

ncase =   2

t	yla	yul	ppm	Tc	eps	pH	r1
1850	0.0000	0.0000	280.0	1.745	8.800	8.222	0.0100
1860	0.0200	0.0008	285.6	1.749	8.898	8.215	0.0100
1870	0.0378	0.0024	290.6	1.753	8.946	8.209	0.0100

1880	0.0555	0.0045	295.5	1.756	9.000	8.203	0.0100
1890	0.0734	0.0073	300.6	1.759	9.058	8.197	0.0100
1900	0.0919	0.0106	305.7	1.763	9.119	8.191	0.0100
1910	0.1112	0.0145	311.1	1.766	9.186	8.185	0.0100
1920	0.1316	0.0189	316.9	1.769	9.258	8.178	0.0100
1930	0.1535	0.0240	323.0	1.773	9.337	8.172	0.0100
1940	0.1769	0.0298	329.5	1.777	9.423	8.164	0.0100
1950	0.2024	0.0362	336.7	1.781	9.517	8.157	0.0100
1960	0.2300	0.0434	344.4	1.785	9.620	8.149	0.0100
1970	0.2602	0.0514	352.9	1.790	9.730	8.140	0.0100
1980	0.2932	0.0602	362.1	1.795	9.848	8.131	0.0100
1990	0.3295	0.0700	372.3	1.800	9.972	8.121	0.0100
2000	0.3693	0.0809	383.4	1.805	10.098	8.111	0.0100
2010	0.4131	0.0929	395.7	1.811	10.220	8.100	0.0100
2020	0.4640	0.1063	409.9	1.817	10.343	8.087	0.0103
2030	0.5252	0.1213	427.1	1.824	10.488	8.072	0.0106
2040	0.5987	0.1385	447.6	1.832	10.673	8.055	0.0108
2050	0.6870	0.1583	472.4	1.842	10.907	8.035	0.0111
2060	0.7935	0.1812	502.2	1.852	11.193	8.013	0.0114
2070	0.9222	0.2081	538.2	1.864	11.536	7.987	0.0117
2080	1.0784	0.2397	582.0	1.876	11.949	7.958	0.0119
2090	1.2688	0.2772	635.3	1.890	12.403	7.925	0.0122
2100	1.5018	0.3220	700.5	1.905	12.921	7.888	0.0125

Atmosphere ppm(2013) =   400.0

GtC Lower Atmosphere(2013) =   724.8

GtC Upper Atmosphere(2013) =   126.6

GtC Long-lived(2013)   = 2498.9

GtC Short-lived(2013)   = 156.7

GtC Upper Layer(2013)   = 987.2

GtC Deep Layer(2013)   = 37227.5

GtC Marine Biota(2013)   =   3.3

ncall =   147

ncase =   3

t	yla	yul	ppm	Tc	eps	pH	r1
1850	0.0000	0.0000	280.0	1.745	8.800	8.222	0.0100
1860	0.0200	0.0008	285.6	1.749	8.898	8.215	0.0100
1870	0.0378	0.0024	290.6	1.753	8.946	8.209	0.0100
1880	0.0555	0.0045	295.5	1.756	9.000	8.203	0.0100
1890	0.0734	0.0073	300.6	1.759	9.058	8.197	0.0100
1900	0.0919	0.0106	305.7	1.763	9.119	8.191	0.0100
1910	0.1112	0.0145	311.1	1.766	9.186	8.185	0.0100
1920	0.1316	0.0189	316.9	1.769	9.258	8.178	0.0100
1930	0.1535	0.0240	323.0	1.773	9.337	8.172	0.0100
1940	0.1769	0.0298	329.5	1.777	9.423	8.164	0.0100
1950	0.2024	0.0362	336.7	1.781	9.517	8.157	0.0100
1960	0.2300	0.0434	344.4	1.785	9.620	8.149	0.0100
1970	0.2602	0.0514	352.9	1.790	9.730	8.140	0.0100
1980	0.2932	0.0602	362.1	1.795	9.848	8.131	0.0100
1990	0.3295	0.0700	372.3	1.800	9.972	8.121	0.0100
2000	0.3693	0.0809	383.4	1.805	10.098	8.111	0.0100
2010	0.4131	0.0929	395.7	1.811	10.220	8.100	0.0100
2020	0.4614	0.1062	409.2	1.817	10.332	8.088	0.0100
2030	0.5146	0.1208	424.1	1.823	10.449	8.075	0.0100
2040	0.5733	0.1370	440.5	1.830	10.587	8.061	0.0100
2050	0.6380	0.1549	458.6	1.837	10.747	8.046	0.0100
2060	0.7094	0.1747	478.6	1.844	10.930	8.031	0.0100
2070	0.7883	0.1965	500.7	1.852	11.134	8.014	0.0100
2080	0.8754	0.2205	525.1	1.859	11.356	7.996	0.0100
2090	0.9715	0.2471	552.0	1.868	11.603	7.978	0.0100
2100	1.0778	0.2765	581.8	1.876	11.871	7.958	0.0100

Atmosphere ppm(2013) =   399.7

GtC Lower Atmosphere(2013) =   724.4

GtC Upper Atmosphere(2013) =   126.6

GtC Long-lived(2013)   = 2498.9

GtC Short-lived(2013)   = 156.6

GtC Upper Layer(2013)   = 987.2

GtC Deep Layer(2013)   = 37227.5

GtC Marine Biota(2013)   =   3.3

ncall =   94

ncase =   4

t	yla	yul	ppm	Tc	eps	pH	r1
1850	0.0000	0.0000	280.0	1.745	8.800	8.222	0.0100
1860	0.0200	0.0008	285.6	1.749	8.898	8.215	0.0100
1870	0.0378	0.0024	290.6	1.753	8.946	8.209	0.0100
1880	0.0555	0.0045	295.5	1.756	9.000	8.203	0.0100
1890	0.0734	0.0073	300.6	1.759	9.058	8.197	0.0100
1900	0.0919	0.0106	305.7	1.763	9.119	8.191	0.0100
1910	0.1112	0.0145	311.1	1.766	9.186	8.185	0.0100
1920	0.1316	0.0189	316.9	1.769	9.258	8.178	0.0100
1930	0.1535	0.0240	323.0	1.773	9.337	8.172	0.0100
1940	0.1769	0.0298	329.5	1.777	9.423	8.164	0.0100
1950	0.2024	0.0362	336.7	1.781	9.517	8.157	0.0100
1960	0.2300	0.0434	344.4	1.785	9.620	8.149	0.0100
1970	0.2602	0.0514	352.9	1.790	9.730	8.140	0.0100
1980	0.2932	0.0602	362.1	1.795	9.848	8.131	0.0100
1990	0.3295	0.0700	372.3	1.800	9.972	8.121	0.0100
2000	0.3693	0.0809	383.4	1.805	10.098	8.111	0.0100
2010	0.4131	0.0929	395.7	1.811	10.220	8.100	0.0100
2020	0.4518	0.1059	406.5	1.815	10.291	8.090	0.0089
2030	0.4789	0.1192	414.1	1.819	10.319	8.083	0.0078
2040	0.4957	0.1323	418.8	1.821	10.332	8.079	0.0067
2050	0.5032	0.1450	420.9	1.822	10.329	8.077	0.0056
2060	0.5028	0.1569	420.8	1.822	10.313	8.077	0.0044
2070	0.4961	0.1678	418.9	1.821	10.285	8.079	0.0033
2080	0.4843	0.1776	415.6	1.819	10.249	8.082	0.0022
2090	0.4690	0.1863	411.3	1.817	10.209	8.086	0.0011
2100	0.4514	0.1938	406.4	1.815	10.167	8.090	0.0000

Atmosphere ppm(2013) =   398.9

GtC Lower Atmosphere(2013) =   723.0

GtC Upper Atmosphere(2013) =   126.5

```
GtC Long-lived(2013)     = 2498.7

GtC Short-lived(2013)    =  156.4

GtC Upper Layer(2013)    =  987.1

GtC Deep Layer(2013)     = 37227.5

GtC Marine Biota(2013)   =    3.3

ncall =  158
```

We can note the following details about this output.

- The sequence of solutions for `ncase=1,2,3,4` is clear. The variation in the model occurs in functions `co2_rate.R`, `co2_rate.m` of Listings 4a and 4b (i.e., the variation in r_1 in Eq. (3.10)).

  R:

```
ncase = 1: r1(1850) = 0.0100,...,r1(2100) = 0.0150
ncase = 2: r1(1850) = 0.0100,...,r1(2100) = 0.0125
ncase = 3: r1(1850) = 0.0100,...,r1(2100) = 0.0100
ncase = 4: r1(1850) = 0.0100,...,r1(2100) = 0.0000
```

  Matlab:

```
ncase = 1: r1(1850) = 0.0100,...,r1(2100) = 0.0150
ncase = 2: r1(1850) = 0.0100,...,r1(2100) = 0.0125
ncase = 3: r1(1850) = 0.0100,...,r1(2100) = 0.0100
ncase = 4: r1(1850) = 0.0100,...,r1(2100) = 0.0000
```

  The variation in $r_1(t)$ is the same for R and Matlab as expected (since these values are defined explictly as a function of t in `co2_rate.R`, `co2_rate.m` (Listings 4a, 4b) and do not depend on the numerical solutions). Also, $r_1(t)$ is the same for $1850 \leq t \leq 2010$, and for $2010 < t \leq 2100$, $r_1(t)$ varies according to the value of `ncase` in Listings 4a and 4b.

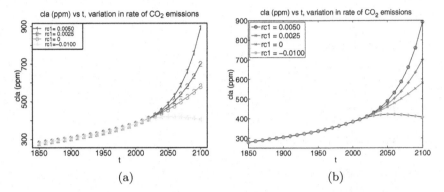

(a)   (b)

Figure 6: (a) Composite plot, $1850 \leq t \leq 2100$ from Listing 2a (R).
(b) Composite plot, $1850 \leq t \leq 2100$ from Listing 2b (Matlab)

- The final values of the atmospheric CO_2 vary as

  R:

  ```
  ncase = 1: ppm(2100) = 892.3
  ncase = 2: ppm(2100) = 700.5
  ncase - 3: ppm(2100) - 581.8
  ncase = 4: ppm(2100) = 406.5
  ```

  Matlab:

  ```
  ncase = 1: ppm(2100) = 892.3
  ncase = 2: ppm(2100) = 700.5
  ncase = 3: ppm(2100) = 581.8
  ncase = 4: ppm(2100) = 406.4
  ```

  These final values are also reflected in Figs. 6a and 6b.
- For `ncase=4` The atmospheric CO_2 peaks at

  R:

  ```
  ppm(2050) = 421.0
  ```

  Matlab:

  ```
  ppm(2050) = 420.9
  ```

  Thus, the model projects CO_2 will go through a maximum, then
  start to decline. This is a hopeful scenario, but whether the

variation in $r_1(t)$ for `ncase=4` in `co2_rate.R`, `co2_rate.m` could be achieved is doubtful.

Although $r_1(t = 2100) = 0$, this does not mean that the emissions source term in Eqs. (3.4a) and (3.10), $Q_c(t)$ is zero. Rather, this source term is constant

$$Q_c(t) = c_1 e^{r_1 t} = c_1 e^{0(2100)} = c_1.$$

Thus, CO_2 continues to enter the lower atmosphere, but at a low enough rate (c_1) that CO_2 moving to other reservoirs (Fig. 1) causes the lower atmosphere *ppm* to go through a maximum, then decline as in Figs. 6a and 6b. A detailed analysis of this effect is available by computing and displaying the individual RHS terms of Eq. (3.4a),

$$\frac{1}{\theta_{ua}^{la}}(y_{ua} - y_{la})$$

$$\frac{1}{\theta_{sb}^{la}}(y_{sb} - y_{la})$$

$$\frac{1}{\theta_{lb}^{la}}(y_{lb} - y_{la})$$

$$\frac{1}{\theta_{ul}^{la}}(y_{ul} - y_{la})$$

$$Q_c(t).$$

These terms can then be added to give the LHS derivative of Eq. (3.4a)

$$\frac{dy_{la}}{dt},$$

which is positive for $1850 \leq t \leq 2050$, then negative for $2050 < t \leq 2100$ as indicated in Figs. 6a and 6b. This type of detailed analysis of the ODE terms gives a complete picture of the dynamics of the model. That is, the model solution is defined by the LHS derivatives of Eqs. (3.3) to (3.9), which in turn, are defined by the RHS terms. As an example, the calculation and display of the RHS terms and the LHS derivative of Eq. (3.4a) is discussed in Appendix A as an illustration of the detailed analysis of the individual terms of a model ODE.

- Also, the ocean pH goes through a minimum (i.e., maximum acidification) at 2050 to 2060.

R:

```
ncase = 4
pH(2040) = 8.079
pH(2050) = 8.077
pH(2060) = 8.077
pH(2070) = 8.079
```

Matlab:

```
ncase = 4
pH(2040) = 8.079
pH(2050) = 8.077
pH(2060) = 8.077
pH(2070) = 8.079
```

This effect could be highlighted by a composite plot added to plot_output.R, plot_output.m of Listings 2a and 2b. This extension is left for the reader, e.g., using plot_out[13],plot_out(13).

- ppm(t=2013) departs slightly from the reported value [12] of 400 because of the variation in r_1 in co2_rate.R, co2_rate.m (Listings 4a and 4b) after t=2010.

R:

```
ncase = 1: ppm(2013) = 400.2
ncase = 2: ppm(2013) = 400.0
ncase = 3: ppm(2013) = 399.7
ncase = 4: ppm(2013) = 398.9
```

Matlab:

```
ncase = 1: ppm(2013) =  400.2
ncase = 2: ppm(2013) =  400.0
ncase = 3: ppm(2013) =  399.7
ncase = 4: ppm(2013) =  398.9
```

- ε can be considered as a type of Henry's law constant relating atmospheric concentration $ppm(t)$ to the equilibrium ocean upper layer total carbon $T_c(t)$. ε increases with atmospheric ppm for ncase=1,2,3,4 in Tables 5a and 5b which indicates that as *ppm* increases, a higher atmospheric *ppm* is required for a given amount of upper layer carbon $T_c(t)$. In other words, uptake of carbon in the ocean upper layer is inhibited as atmospheric CO_2 increases suggesting a limit to ocean uptake of CO_2 (so that the oceans are not effectively an infinite CO_2 reservoir as generally perceived).

• Computationally, Matlab is more efficient than R, but in both cases, the computational effort is quite modest.

R:

```
ncase = 1: ncall = 363
ncase = 2: ncall = 362
ncase = 3: ncall = 210
ncase = 4: ncall = 489
```

Matlab:

```
ncase = 1: ncall = 153
ncase = 2: ncall = 147
ncase = 3: ncall =  94
ncase = 4: ncall = 158
```

In summary, the scenarios of ncase=1,2,3,4 move the output for $t > 2010$ from the base case (ncase = 1) through successively decreasing CO_2 emission rates (lower r_1 as programmed in co2_rate.R, co2_rate.m of Listings 4a and 4b). Whether these lower emissions rates can actually be achieved remains to be seen, but the model output projects the corresponding reduction of atmospheric CO_2 concentrations. Similar conclusions follow for ocean acidification.

7.2. Comparison with Observational Data

The increase in atmospheric CO_2 is reflected in the famous CO_2 plot from Mauna Loa, Hawaii [8] in Fig. 7.

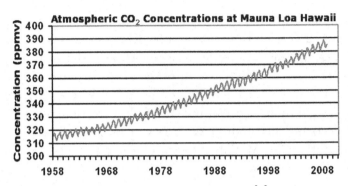

Figure 7: Mauna Loa CO_2 concentration vs t [8]
Source: Oak Ridge National Laboratory.

The saw tooth character of this plot is due to annual fluctuations but the overall trend of increasing CO_2 is clear (the milestone point $c_{la}(t = 2013) = 400$ will eventually be included in the plot). The model output can be compared with Fig. 7 by adding points from Fig. 7 to plot_output.R, plot_output.m of Listings 2a and 2b. For Listing 2a (in R), plot_out[1]==1, the observed data from Fig. 7 can be added as follows.

```
plot_output=function(plot_out){
#
# Select plots
#
# c_{la}(ppm) vs t
  if(plot_out[1]==1){
  par(mfrow=c(1,1))
  plot(tout,ppm,xlab="t",ylab="cla(ppm)",type="l",lwd=2,
      main="cla(ppm) vs t\n solid - model, 0 - Mauna Loa");
  tml=rep(0,6);
  cml=rep(0,6);
  tml[1]=1960; cml[1]=315;
  tml[2]=1970; cml[2]=325;
  tml[3]=1980; cml[3]=336;
  tml[4]=1990; cml[4]=351;
  tml[5]=2000; cml[5]=366;
  tml[6]=2010; cml[6]=385;
  points(tml,cml,pch="0",lwd=2);
  }
```

We can note the following details about this code.

- The `plot` title `main="cla(ppm) vs t\n solid - model, 0 -` `Mauna Loa"` indicates a comparison of the model and the Mauna Loa data.
- Two vectors, `tml`, `cml`, are declared for six observational data (`ml` is used to designate Mauna Loa).

```
tml=rep(0,6);
cml=rep(0,6);
```

- Six data pairs are added from Fig. 7.

```
tml[1]=1960; cml[1]=315;
tml[2]=1970; cml[2]=325;
tml[3]=1980; cml[3]=336;
tml[4]=1990; cml[4]=351;
tml[5]=2000; cml[5]=366;
tml[6]=2010; cml[6]=385;
```

- The six data pairs are added to the plot with the `points` utility.

```
points(tml,cml,pch="0",lwd=2);
```

The graphical output from Listings 1a to 5a (for the R routines) with 2a modified as above gives Fig. 8 (for `plot_out[1]=1, ncase=1` in Listing 1a). We can note the following details about Fig. 8.

- The model generally overestimates (exceeds) the data. This difference in not unexpected since the emission rate over the interval $1850 \leq t \leq 2010$ is expressed by a single exponential (from Eq. (3.10)), $Q_c = 4.92 \times 10^{-3} e^{0.01t}$ (see `co2_rate.R` of Listing 4a), and it is unlikely that this single function could represent the observed data of Fig. 7 over $1960 \leq t \leq 2010$.
- The misfit between the model and data (model > data) appears to occur from the beginning, $t = 1850$. This suggests that Q_c should be lower at early times, then increased later so that $c_{la}(t = 2013) = 400$ is maintained. This in turn suggests that c_1, r_1 in Eq. (3.10) could be programmed as a function of t (rather than maintain the constant values $c_1 = 4.92 \times 10^{-3}, r_1 = 0.01$) to achieve better

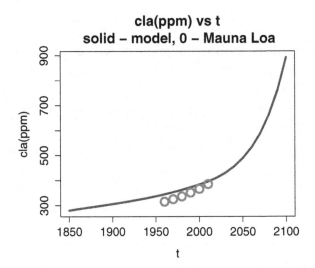

Figure 8: Comparison of model output with Mauna Loa data, ncase=1

agreement between the model and observations. This variation of the model is discussed next.

- This requirement of lower initial emission rates seems intuitively reasonable since shortly after $t = 1850$, the CO_2 emission rates were low, then increased as industrialization expanded.

The following extension of co2_rate.R of Listing 4a provides a t variation of c_1, r_1.

Listing 4c: Function co2_rate.R with t variation of c_1, r_1

```
CO2_rate=function(t,ncase){
#
# Function CO2_rate returns the constants c1, r1 in the CO2
# source term
#
#    CO2_rate = c1*exp(r1*(t-1850))
#
# for the case ncase.
#
# Set c1, r1
   r1c=0;
```

```
if(t<=1960){r1b=0.0;c1=5.0e-03};
if((t>1960)&(t<=2000)){r1b=0.0085;c1=6.2e-03};
if(t >2000){r1b=0.010;c1=7.5e-03};
#
# Change r1 for t > 2010
  if(t>2010){
    if(ncase==1){r1c= 0.0050;}
    if(ncase==2){r1c= 0.0025;}
    if(ncase==3){r1c= 0.0000;}
    if(ncase==4){r1c=-0.0100;}
  }
#
# Linear interpolation in t between 2010 and 2100
  r1=r1b+r1c*(t-2010)/(2100-2010);
#
# Return c1,r1
  c1 <<- c1;
  r1 <<- r1;
#
# End of CO2_rate
  }
```

The only change in `co2_rate.R` in Listing 4a is the programming of c_1, r_1 as a function of t.

```
#
# Set c1, r1
  r1c=0;
  if(t<=1960){r1b=0.0;c1=5.0e-03};
  if((t>1960)&(t<=2000)){r1b=0.0085;c1=6.2e-03};
  if(t >2000){r1b=0.010;c1=7.5e-03};
```

The values for `c1` and `r1b` were determined by trial and error to achieve better agreement (than Fig. 8) between the model and the observed data of Fig. 7. Figure 9 is the resulting plot.

The preceding modification for c_1, r_1 as a function of t can be executed via the main program of Listing 1a to give the agreement between the model and data as in Fig. 9.

To conclude, the preceding discussion is in terms of R. The corresponding programming in Matlab is left as an exercise for the reader

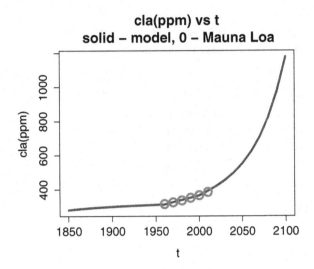

Figure 9: Comparison of model output with Mauna Loa data, ncase=1, time adjusted c_1, r_1

(see Appendix E). This includes the extension of the plot_out(1)==1 programming in plot_output.R of Listing 2a (to include the observational data from Fig. 7), and the modification of the programming in model_1.R of Listing 3a (to vary c_1, r_1 as a function of t).

8 SOURCES OF INFORMATION

The literature pertaining to CO_2 is very large and increasing rapidly. Thus, because a bibliography would soon become out of date, we provide an updated file with general sources of information and references for further reading on request from wes1@lehigh.edu. Here, we provide just the links to some sources of information, e.g., [8].

- [1] American Meteorological Society

  http://journals.ametsoc.org/loi/bams

- [5] Center for Climate Change Communication

  http://www.climatechangecommunication.org

- [6] Center for Climate and Energy Solutions, China

  http://www.c2es.org/international/key-country-policies/china

- [7] Center for International Climate and Environment Research

  http://www.cicero.uio.no/home/index_e.aspx

- [8] DOE/ORNL Carbon Dioxide Information Analysis Center

  http://cdiac.esd.ornl.gov
  http://cdiac.ornl.gov/ftp/ndp030/global.1751_2010.ems

- [10] Earth System Research Laboratory

  http://www.esrl.noaa.gov/research/themes/carbon/

- [11] Environmental Protection Agency

  http://www.epa.gov/climatechange/

- [13] Global Carbon Project

  http://www.globalcarbonproject.org/index.htm
  http://www.globalcarbonproject.org/carbontrends

- [16] National Center for Atmospheric Research (NCAR)

  https://ncar.ucar.edu
  https://ncar.ucar.edu/learn-more-about/climate

- [17] National Climate Assessment

  http://nca2014.globalchange.gov

- [18] National Climate Data Center

  http://www.ncdc.noaa.gov/

- [19] National Operational Hydrologic Remote Sensing Center

  http://www.nohrsc.noaa.gov/nsa

- [20] National Snow and Ice Data Center

  http://nsidc.org/noaa

- [22] Royal Meteorological Society

  http://www.rmets.org/

- [28] Tyndall Centre for Climate Change Research

  http://www.tyndall.ac.uk/corinne-le-quere

- [29] World Meteorological Organization

  https://www.wmo.int

- [30] World Wildlife Fund

  http://www.worldwildlife.org/

Other sources include Journals such as *Nature, Science, Climate Dynamics, International Journal of Climatology.*

9 SUMMARY AND CONCLUSIONS

Our intent in presenting the CO_2 model and output is generally to identify some key numbers that give a quantitative perspective of the global CO_2 dynamics. Of particular interest is the evolution of these numbers with time, e.g., atmospheric CO_2 expressed in ppm and ocean pH.

Extensions of the model can also be considered. For example, as the oceans warm, the chemistry discussed previously will be affected. Specifically, variations in the equilibrium constants of reactions (r1) to (r5) with temperature could be investigated using equations in [2].

The authors would welcome knowing about experiences in using the model and suggestions for improvements (directed to wes1@lehigh.edu).

Acknowledgments

This work on CO_2 modeling and computer-based analysis was initiated by the ECOMSETS grant from the National Science Foundation, 1976.

REFERENCES

[1] American Meteorological Society
http://journals.ametsoc.org/loi/bams

[2] Bacastow, R., and C. D. Keeling (1972), *Atmospheric Carbon Dioxide and Radiocarbon in the Natural Carbon Cycle: II Changes from A.D. 1700 to 2070 as Deduced from a Geochemical Model*, in Carbon and the Biosphere (G. M. Woodwell and E. V. Pecan, eds.), Proceedings of the 24th Brookhaven Symposium in Biology, Upton, NY, May 16–18; published by the Technical Information Center, Office of Information Services, United States Atomic Energy Commission.

[3] Broecker W. S., and T.-H. Peng (1982), *Tracers in the Sea*, Lamont-Doherty Earth Observatory.

[4] Canadell, J. G., *et al.* (2007), Contributions to accelerating atmospheric CO_2 growth from economic activity, carbon intensity, and efficiency of natural sinks, *PNAS*, **104**(47), 18866–18870.

[5] Center for Climate Change Communication
http://www.climatechangecommunication.org

[6] Center for Climate and Energy Solutions, China
http://www.c2es.org/international/key-country-policies/china

[7] Center for International Climate and Environment Research
http://www.cicero.uio.no/home/index_e.aspx

[8] DOE/ORNL Carbon Dioxide Information Analysis Center
http://cdiac.esd.ornl.gov
http://cdiac.ornl.gov/ftp/ndp030/global.1751_2010.ems

[9] Doney, S. C. (2006), The Dangers of Ocean Acidification, *Scientific American*, March, 58–65.

[10] Earth System Research Laboratory
http://www.esrl.noaa.gov/research/themes/carbon/

[11] Environmental Protection Agency
http://www.epa.gov/climatechange/

[12] Gillis, J., *New York Times*, May 10, 2013, indicating the atmospheric CO_2 concentration has passed 400 ppm. Available online at
http://www.nytimes.com/2013/05/11/science/earth/carbon-dioxide-level-passes-long-feared-milestone.html?hp&_r=0
A series of articles pertaining to climate change, global warming and CO_2 is also listed here. A related graphic is available at
http://www.nytimes.com/interactive/2013/05/10/science/crossing-a-line.html?ref=earth

[13] Global Carbon Project
http://www.globalcarbonproject.org/index.htm
http://www.globalcarbonproject.org/carbontrends

[14] Lifton, R.J. (2014), *New York Times*, August 23, 2014, Discussion of increasing general awareness of climate change.

[15] Millero, F.J. (2006), *Chemical Oceanography*, 3rd edn., Taylor and Francis/CRC, Boca Raton, FL.

[16] National Center for Atmospheric Research (NCAR)
https://ncar.ucar.edu
https://ncar.ucar.edu/learn-more-about/climate

[17] National Climate Assessment
http://nca2014.globalchange.gov

[18] National Climate Data Center
http://www.ncdc.noaa.gov/

[19] National Operational Hydrologic Remote Sensing Center
http://www.nohrsc.noaa.gov/nsa

[20] National Snow and Ice Data Center
http://nsidc.org/noaa

[21] Pilson, M.E.Q. (1998), *An Introduction to the Chemistry of the Sea*, Prentice Hall, Upper Saddle River, NJ

[22] Royal Meteorological Society
http://www.rmets.org/

[23] Revelle, R., and H.E. Suess (1957), Carbon dioxide exchange between atmosphere and ocean and the question of an increase of atmospheric CO2 during past decades, *Tellus*, 9, 18–27.

[24] Sarmiento, J.L., and N. Gruber (2002), Sinks for anthropogenic carbon, *Physics Today*, 55 (8), 30–36.

[25] Sarmiento, J.L., and N. Gruber (2006), *Ocean Biogeochemical Dynamics*, Princeton University Press, Princeton, NJ.

[26] Soetaert, K., J. Cash, and F. Mazzia (2012), *Solving Differential Equations in R*, Springer-Verlag, Heidelberg, Germany.

[27] Tomizuka, A. (2009), Is a box model effective for understanding the carbon cycle?, *American Journal of Physics*, **77** (2), 156–163.

[28] Tyndall Centre for Climate Change Research

http://www.tyndall.ac.uk/corinne-le-quere

[29] World Meteorological Organization

https://www.wmo.int

[30] World Wildlife Fund

http://www.worldwildlife.org/

APPENDIX A: CALCULATION OF ODE TERMS

The solution of an ODE model is determined by the LHS derivatives of the ODEs which are integrated numerically, for example, by `ode` in Listings 1a and 1c and `lsodes` in Listings 1b and 1d. The derivatives in turn are defined by the RHS terms of the ODEs. Thus, in order to obtain a detailed picture of the model solution, the ODE LHS and RHS terms can be examined in detail. This type of analysis is illustrated by the following application to Eq. (3.4a).

First a modification of Listings 1c and 1d follows in which the plotting at the end is replaced with plotting of the LHS and RHS terms of Eq. (3.4a).

Listing 1e: Revised main program in R with calculation and plotting of the LHS, RHS of Eq. (3.4a)

R:

```
#
# Delete previous workspaces
  rm(list=ls(all=TRUE))
#
# Access ODE integrator
  library("deSolve");
#
# Access files
  setwd("c:/model_R");
  source("model_1.R");
  source("splines.R");
  source("CO2_rate.R");
```

```
#
# Declare (preallocate) arrays
  nout=26;
   r1e=rep(0,nout);  ppm=rep(0,nout); eps=rep(0,nout);
    Tc=rep(0,nout); hion=rep(0,nout);  pH=rep(0,nout);
   co2=rep(0,nout); hco3=rep(0,nout); co3=rep(0,nout);
  boh4=rep(0,nout);
#
# Dimensional ODE dependent variables
  cla=rep(0,nout);cua=rep(0,nout);csb=rep(0,nout);
  clb=rep(0,nout);cul=rep(0,nout);cdl=rep(0,nout);
  cmb=rep(0,nout);
#
# Select case (see CO2_rate.R)
  ncase=1;
#
# Equilibrium constants (in millimoles/lt units)
  k0=0.03347e-03;
  k1=9.747e-04;
  k2=8.501e-07;
  kb=1.881e-06;
  kw=6.46e-09;
#
# Boron
  B=0.409;
#
# Initial conditions
  n=7;
  y0=rep(0,n);
#
# Independent variable for ODE integration
  t0=1850;tf=2100;
  tout=seq(from=t0,to=tf,by=(tf-t0)/(nout-1));
  ncall=0;
#
# Initialized carbon reservoirs (GtC)
  c0la=0.85*597;
  c0ua=0.15*597;
  c0sb=110;
  c0lb=2190;
  c0ul=900;
  c0dl=37100;
```

```
   c0mb=3;
#
# ODE integration
   out=ode(func=model_1,times=tout,y=y0);
#
# Set up spline interpolation
   ppm0=200;dppm=100;max=16;
   splines(ppm0,dppm,max);
#
# Display selected output
   cat(sprintf("\n ncase = %4d\n",ncase));
   cat(sprintf(
   "\n    t      yla     yul     ppm     Tc      eps      pH
   r1\n"));
   for(it in 1:nout){
#
# CO2 emissions rate (in lower atmosphere)
   CO2_rate(tout[it],ncase);
   r1e[it]=r1;
#
# CO2 ppm (in lower atmosphere)
   ppm[it]=280*(1+out[it,2]);
#
# Total carbon
   Tc[it]=Tcs(ppm[it]);
#
# Evasion factor
   if(it==1){eps[1]=8.8;}
   if(it >1){
     num=(ppm[it]-ppm[it-1])/ppm[it-1];
     den=( Tc[it]- Tc[it-1])/ Tc[it-1];
     eps[it]=num/den;
   }
#
# Hydrogen ion
   hion[it]=hions(ppm[it]);
#
# pH
   pH[it]=pHs(ppm[it]);
#
# CO2 [CO2 + H2CO3 in upper layer)
   co2[it]=k0*ppm[it];
```

```
#
# Bicarbonate
  hco3[it]=k0*k1*ppm[it]/hion[it];
#
# Carbonate
  co3[it]=k0*k1*k2*ppm[it]/hion[it]^2;
#
# Boron
  boh4[it]=B/(1+hion[it]/kb);
#
# Selected output
  cat(sprintf("%5.0f%8.4f%8.4f%8.1f%8.3f%8.3f%8.3f%8.4f\n",
    out[it,1],out[it,2],out[it,6],ppm[it],Tc[it],eps[it],
    pH[it],r1e[it]));
#
# CO2 GtC (in all reservoirs) vs t
  cla[it]=c0la*(1+out[it,2]);
  cua[it]=c0ua*(1+out[it,3]);
  csb[it]=c0sb*(1+out[it,4]);
  clb[it]=c0lb*(1+out[it,5]);
  cul[it]=c0ul*(1+out[it,6]);
  cdl[it]=c0dl*(1+out[it,7]);
  cmb[it]=c0mb*(1+out[it,8]);
  }
#
# CO2 ppm at 2013 (by linear interpolation between 2010
# and 2020)
  p2013=ppm[17]+(ppm[18]-ppm[17])*(2013-2010)/(2020-2010);
#
# GtC (in all reservoirs) at 2013
  cla2013=cla[17]+(cla[18]-cla[17])*(2013-2010)/(2020-2010);
  cua2013=cua[17]+(cua[18]-cua[17])*(2013-2010)/(2020-2010);
  clb2013=clb[17]+(clb[18]-clb[17])*(2013-2010)/(2020-2010);
  csb2013=csb[17]+(csb[18]-csb[17])*(2013-2010)/(2020-2010);
  cul2013=cul[17]+(cul[18]-cul[17])*(2013-2010)/(2020-2010);
  cdl2013=cdl[17]+(cdl[18]-cdl[17])*(2013-2010)/(2020-2010);
  cmb2013=cmb[17]+(cmb[18]-cmb[17])*(2013-2010)/(2020-2010);
#
# GtC at 2013 output
  cat(sprintf("\n\n Atmosphere ppm(2013) = %6.1f\n",p2013));
  cat(sprintf("\n GtC Lower Atmosphere(2013) = %6.1f\n",
    cla2013));
```

```
  cat(sprintf("\n GtC Upper Atmosphere(2013) = %6.1f\n",
    cua2013));
  cat(sprintf("\n GtC Long-Lived(2013)        = %6.1f\n",
    clb2013));
  cat(sprintf("\n GtC Short-Lived(2013)       = %6.1f\n",
    csb2013));
  cat(sprintf("\n GtC Upper Layer(2013)       = %6.1f\n",
    cul2013));
  cat(sprintf("\n GtC Deep Layer(2013)        = %6.1f\n",
    cdl2013));
  cat(sprintf("\n GtC Marine Biota(2013)      = %6.1f\n",
    cmb2013));
  cat(sprintf("\n ncall = %4d\n",ncall));
#
# CO2 sensitivity (C^o/ppm CO2)
  co2s=0.005;
#
# Compute and plot derivatives, RHS terms
#
# Model dependent variables
  yla=rep(0,nout);yua=rep(0,nout);ysb=rep(0,nout);
  ylb=rep(0,nout);yul=rep(0,nout);ydl=rep(0,nout);
  ymb=rep(0,nout);
  for(it in 1:nout){
    yla[it]=out[it,2];yua[it]=out[it,3];
    ysb[it]=out[it,4];ylb[it]=out[it,5];
    yul[it]=out[it,6];ydl[it]=out[it,7];
    ymb[it]=out[it,8];
  }
#
# Model dependent variable derivatives
  dyla=rep(0,nout);dyua=rep(0,nout);dysb=rep(0,nout);
  dylb=rep(0,nout);dyul=rep(0,nout);dydl=rep(0,nout);
  dymb=rep(0,nout);
  for(it in 1:nout){
    CO2_rate(tout[it],ncase);
    dyla[it]=    1/5*(yua[it]-yla[it])+
              1/0.75*(ysb[it]-yla[it])+
               1/150*(ylb[it]-yla[it])+
                1/30*(yul[it]-yla[it])+
               c1*exp(r1*(tout[it]-1850));
    dyua[it]=    1/3*(yla[it]-yua[it]);
```

```
    dysb[it]=  1/0.75*(yla[it]-ysb[it]);
    dylb[it]=  1/150*(yla[it]-ylb[it]);
    dyul[it]=   1/80*(yla[it]-yul[it])+
                1/200*(ydl[it]-yul[it])+
                1/5*(ymb[it]-yul[it]);
    dydl[it]=  1/1500*(yul[it]-ydl[it]);
    dymb[it]=  1/10*(yul[it]-ymb[it]);
  }
#
# Plot numerical solutions, derivatives
#
# yla, dyla/dt
  par(mfrow=c(1,2))
  plot(tout,yla,xlab="t (year)",ylab="yla(t)",
     xlim=c(1850,2100),main="yla",type="l",lwd=2);
  plot(tout,dyla,xlab="t (year)",ylab="dyla(t)/dt",
     xlim=c(1850,2100),main="dyla/dt",type="l",lwd=2);
#
# yua, dyua/dt
  par(mfrow=c(1,2))
  plot(tout,yua,xlab="t (year)",ylab="yua(t)",
     xlim=c(1850,2100),main="yua",type="l",lwd=2);
  plot(tout,dyua,xlab="t (year)",ylab="dyua(t)/dt",
     xlim=c(1850,2100),main="dyua/dt",type="l",lwd=2);
#
# ysb, dysb/dt
  par(mfrow=c(1,2))
  plot(tout,ysb,xlab="t (year)",ylab="ysb(t)",
     xlim=c(1850,2100),main="ysb",type="l",lwd=2);
  plot(tout,dysb,xlab="t (year)",ylab="dysb(t)/dt",
     xlim=c(1850,2100),main="dysb/dt",type="l",lwd=2);
#
# ylb, dylb/dt
  par(mfrow=c(1,2))
  plot(tout,ylb,xlab="t (year)",ylab="ylb(t)",
     xlim=c(1850,2100),main="ylb",type="l",lwd=2);
  plot(tout,dylb,xlab="t (year)",ylab="dylb(t)/dt",
     xlim=c(1850,2100),main="dylb/dt",type="l",lwd=2);
#
# yul, dyul/dt
  par(mfrow=c(1,2))
  plot(tout,yul,xlab="t (year)",ylab="yul(t)",
```

```
      xlim=c(1850,2100),main="yul",type="l",lwd=2);
  plot(tout,dyul,xlab="t (year)",ylab="dyul(t)/dt",
      xlim=c(1850,2100),main="dyul/dt",type="l",lwd=2);
#
# ydl, dydl/dt
  par(mfrow=c(1,2))
  plot(tout,ydl,xlab="t (year)",ylab="ydl(t)",
      xlim=c(1850,2100),main="ydl",type="l",lwd=2);
  plot(tout,dydl,xlab="t (year)",ylab="dydl(t)/dt",
      xlim=c(1850,2100),main="dydl/dt",type="l",lwd=2);
#
# ymb, dymb/dt
  par(mfrow=c(1,2))
  plot(tout,ymb,xlab="t (year)",ylab="ymb(t)",
      xlim=c(1850,2100),main="ymb",type="l",lwd=2);
  plot(tout,dymb,xlab="t (year)",ylab="dymb(t)/dt",
      xlim=c(1850,2100),main="dymb/dt",type="l",lwd=2);
#
# RHS terms
  term1=rep(0,nout);term2=rep(0,nout);term3=rep(0,nout);
  term4=rep(0,nout);term5=rep(0,nout);
  for(it in 1:nout){
    CO2_rate(tout[it],ncase);
    term1[it]=    1/5*(yua[it]-yla[it]);
    term2[it]= 1/0.75*(ysb[it]-yla[it]);
    term3[it]=  1/150*(ylb[it]-yla[it]);
    term4[it]=   1/30*(yul[it]-yla[it]);
    term5[it]=  c1*exp(r1*(tout[it]-1850));
  }
#
# Numerical output
  cat(sprintf("\n           t                             "));
  cat(sprintf("\n      term1     term2      temp3     term4"));
  cat(sprintf("\n      term5  dyla/dt      yla             "));
  for(it in 1:nout){
    cat(sprintf("\n\n t = %5.0f",tout[it]));
    cat(sprintf("\n%10.4f%10.4f%10.4f%10.4f",
        term1[it],term2[it],term3[it],term4[it]));
    cat(sprintf("\n%10.4f%10.4f%10.4f",
        term5[it],dyla[it],yla[it]));
  }
#
```

```
# Graphical output
  par(mfrow=c(3,2))
  plot(tout,term1,xlab="t (year)",ylab="term1",
      xlim=c(1850,2100),ylim=c(-0.1,0.1),
      main="1/5*(yua(t)-yla(t))",type="l",lwd=2);
  plot(tout,term2,xlab="t (year)",ylab="term2",
      xlim=c(1850,2100),ylim=c(-0.1,0.1),
      main="1/0.75*(ysb(t)-yla(t))",type="l",lwd=2);
  plot(tout,term3,xlab="t (year)",ylab="term3",
      xlim=c(1850,2100),ylim=c(-0.1,0.1),
      main="1/150*(ylb[it]-yla[it])",type="l",lwd=2);
  plot(tout,term4,xlab="t (year)",ylab="term4",
      xlim=c(1850,2100),ylim=c(-0.1,0.1),
      main="1/30*(yul(t)-yla(t))",type="l",lwd=2);
  plot(tout,term5,xlab="t (year)",ylab="term5",
      xlim=c(1850,2100),ylim=c(-0.1,0.2),
      main="c1*exp(r1*(t-1850))",type="l",lwd=2);
  plot(tout,dyla,xlab="t (year)",ylab="dla(t)/dt",
      xlim=c(1850,2100),ylim=c(-0.1,0.1),
      main="dla(t)/dt",type="l",lwd=2);
```

Listing 1f: Revised main program in Matlab with calculation and plotting of the LHS, RHS of Eq. (3.4a)

Matlab:

```
%
% Clear previous files
  clear all
  clc
%
% General parameters (shared with other routines)
  global ncall ncase
%
% Spline tables
  global Tc hion pH
%
% Variables for plotting
  global t      y     ppm    co2  hco3    co3  boh4...
         cla   cua    csb    clb   cul    cdl   cmb...
         co2s  eps
%
% Equilibrium constants, D
```

```
  global k0 k1 k2 kb kw B
%
% Select case (see CO2_rate.m)
  ncase=1;
%
% Equilibrium constants (in millimoles/lt units)
  k0=0.03347e-03;
  k1=9.747e-04;
  k2=8.501e-07;
  kb=1.881e-06;
  kw=6.46e-09;
%
% Boron
  B=0.409;
%
% Initial conditions
  n=7;
  y0=zeros(1,n);
%
% Independent variable for ODE integration
  t0=1850;tf=2100;nout=26;
  tout=[t0:10:tf]';
  ncall=0;
%
% Initialized carbon reservoirs (GtC)
  c0la=0.85*597;
  c0ua=0.15*597;
  c0sb=110;
  c0lb=2190;
  c0ul=900;
  c0dl=37100;
  c0mb=3;
%
% ODE integration
  reltol=1.0e-06;abstol=1.0e-06;
  options=odeset('RelTol',reltol,'AbsTol',abstol);
  [t,y]=ode15s(@model_1,tout,y0,options);
%
% Set up spline interpolation
  ppm0=200;dppm=100;max=16;
  [Tcs,hions,pHs]=splines(ppm0,dppm,max);
%
```

```
% Display selected output
  fprintf('\n ncase = %4d\n',ncase);
  fprintf(
  '\n    t       yla     yul     ppm     Tc      eps     pH
  r1\n')
  for it=1:nout
%
%    CO2 emissions rate (in lower atmosphere)
     [c1,r1]=CO2_rate(t(it));
     r1_t(it)=r1;
%
%    CO2 ppm (in lower atmosphere)
     ppm(it)=280*(1+y(it,1));
%
% Total carbon
  Tci(it)=ppval(Tcs,ppm(it));
%
% Evasion factor
  if(it==1)eps(1)=8.8;end
  if(it >1)
    num=(ppm(it)-ppm(it-1))/ppm(it-1);
    den=(Tci(it)-Tci(it-1))/Tci(it-1);
    eps(it)=num/den;
  end
%
%    Hydrogen ion
     hion(it)=ppval(hions,ppm(it));
%
%    pH
     pH(it)=ppval(pHs,ppm(it));
%
%    CO2 (CO2 + H2CO3 in upper layer)
     co2(it)=k0*ppm(it);
%
%    Bicarbonate
     hco3(it)=k0*k1*ppm(it)/hion(it);
%
%    Carbonate
     co3(it)=k0*k1*k2*ppm(it)/hion(it)^2;
%
%    Boron
     boh4(it)=B/(1+hion(it)/kb);
```

```
%
%   Selected output
    fprintf('%5.0f%8.4f%8.4f%8.1f%8.3f%8.3f%8.3f%8.4f\n',...
      t(it),y(it,1),y(it,5),ppm(it),Tc(it),eps(it),pH(it),
      r1_t(it));
%
%   CO2 GtC (in all reservoirs) vs t
    cla(it)=c0la*(1+y(it,1));
    cua(it)=c0ua*(1+y(it,2));
    csb(it)=c0sb*(1+y(it,3));
    clb(it)=c0lb*(1+y(it,4));
    cul(it)=c0ul*(1+y(it,5));
    cdl(it)=c0dl*(1+y(it,6));
    cmb(it)=c0mb*(1+y(it,7));
    end
%
%   CO2 ppm at 2013 (by linear interpolation between 2010
%   and 2020)
    p2013=ppm(17)+(ppm(18)-ppm(17))*(2013-2010)/(2020-2010);
%
%   GtC (in all reservoirs) at 2013
    cla2013=cla(17)+(cla(18)-cla(17))*(2013-2010)/(2020-2010);
    cua2013=cua(17)+(cua(18)-cua(17))*(2013-2010)/(2020-2010);
    clb2013=clb(17)+(clb(18)-clb(17))*(2013-2010)/(2020-2010);
    csb2013=csb(17)+(csb(18)-csb(17))*(2013-2010)/(2020-2010);
    cul2013=cul(17)+(cul(18)-cul(17))*(2013-2010)/(2020-2010);
    cdl2013=cdl(17)+(cdl(18)-cdl(17))*(2013-2010)/(2020-2010);
    cmb2013=cmb(17)+(cmb(18)-cmb(17))*(2013-2010)/(2020-2010);
%
%   GtC at 2013 output
    fprintf('\n\n Atmosphere ppm(2013) = %6.1f\n',p2013);
    fprintf('\n GtC Lower Atmosphere(2013) = %6.1f\n',cla2013);
    fprintf('\n GtC Upper Atmosphere(2013) = %6.1f\n',cua2013);
    fprintf('\n GtC Long-lived(2013)       = %6.1f\n',clb2013);
    fprintf('\n GtC Short-lived(2013)      = %6.1f\n',csb2013);
    fprintf('\n GtC Upper Layer(2013)      = %6.1f\n',cul2013);
    fprintf('\n GtC Deep Layer(2013)       = %6.1f\n',cdl2013);
    fprintf('\n GtC Marine Biota(2013)     = %6.1f\n',cmb2013);
    fprintf('\n ncall = %4d\n',ncall);
%
%   CO2 sensitivity (C^o/ppm CO2)
    co2s=0.005;
```

```
%
% Compute and plot derivatives, RHS terms
%
% Model dependent variables
  for(it=1:nout)
    yla(it)=y(it,1);yua(it)=y(it,2);
    ysb(it)=y(it,3);ylb(it)=y(it,4);
    yul(it)=y(it,5);ydl(it)=y(it,6);
    ymb(it)=y(it,7);
  end
%
% Model dependent variable derivatives
  for(it=1:nout)
    [c1,r1]=CO2_rate(tout(it));
    dyla(it)=     1/5*(yua(it)-yla(it))+...
                  1/0.75*(ysb(it)-yla(it))+...
                  1/150*(ylb(it)-yla(it))+...
                  1/30*(yul(it)-yla(it))+...
                  c1*exp(r1*(tout(it)-1850));
    dyua(it)=     1/3*(yla(it)-yua(it));
    dysb(it)=  1/0.75*(yla(it)-ysb(it));
    dylb(it)=   1/150*(yla(it)-ylb(it));
    dyul(it)=    1/80*(yla(it)-yul(it))+...
                 1/200*(ydl(it)-yul(it))+...
                 1/5*(ymb(it)-yul(it));
    dydl(it)=  1/1500*(yul(it)-ydl(it));
    dymb(it)=   1/10*(yul(it)-ymb(it));
  end
%
% Plot numerical solutions, derivatives
%
% yla, dyla/dt
  figure(1);
  subplot(1,2,1);
  plot(tout,yla,'-');
  xlabel('t (year)');ylabel('yla(t)');
  title('yla');axis tight;
  subplot(1,2,2);
  plot(tout,dyla,'-');
  xlabel('t (year)');ylabel('dyla(t)/dt');
  title('dyla/dt');axis tight;
%
```

```
% yua, dyua/dt
  figure(2);
  subplot(1,2,1);
  plot(tout,yua,'-');
  xlabel('t (year)');ylabel('yua(t)');
  title('yua');axis tight;
  subplot(1,2,2);
  plot(tout,dyua,'-');
  xlabel('t (year)');ylabel('dyua(t)/dt');
  title('dyua/dt');axis tight;
%
% ysb, dysb/dt
  figure(3);
  subplot(1,2,1);
  plot(tout,ysb,'-');
  xlabel('t (year)');ylabel('ysb(t)');
  title('ysb');axis tight;
  subplot(1,2,2);
  plot(tout,dysb,'-');
  xlabel('t (year)');ylabel('dysb(t)/dt');
  title('dysb/dt');axis tight;
%
% ylb, dylb/dt
  figure(4);
  subplot(1,2,1);
  plot(tout,ylb,'-');
  xlabel('t (year)');ylabel('ylb(t)');
  title('ylb');axis tight;
  subplot(1,2,2);
  plot(tout,dylb,'-');
  xlabel('t (year)');ylabel('dylb(t)/dt');
  title('dylb/dt');axis tight;
%
% yul, dyul/dt
  figure(5);
  subplot(1,2,1);
  plot(tout,yul,'-');
  xlabel('t (year)');ylabel('yul(t)');
  title('yul');axis tight;
  subplot(1,2,2);
  plot(tout,dyul,'-');
  xlabel('t (year)');ylabel('dyul(t)/dt');
```

```
    title('dyul/dt');axis tight;
%
% ydl, dydl/dt
    figure(6);
    subplot(1,2,1);
    plot(tout,ydl,'-');
    xlabel('t (year)');ylabel('ydl(t)');
    title('ydl');axis tight;
    subplot(1,2,2);
    plot(tout,dydl,'-');
    xlabel('t (year)');ylabel('dydl(t)/dt');
    title('dydl/dt');axis tight;
%
% ymb, dymb/dt
    figure(7);
    subplot(1,2,1);
    plot(tout,ymb,'-');
    xlabel('t (year)');ylabel('ymb(t)');
    title('ymb');axis tight;
    subplot(1,2,2);
    plot(tout,dymb,'-');
    xlabel('t (year)');ylabel('dymb(t)/dt');
    title('dymb/dt');axis tight;
%
% RHS terms
    for(it=1:nout)
      [c1,r1]=CO2_rate(tout(it));
      term1(it)=    1/5*(yua(it)-yla(it));
      term2(it)= 1/0.75*(ysb(it)-yla(it));
      term3(it)=  1/150*(ylb(it)-yla(it));
      term4(it)=   1/30*(yul(it)-yla(it));
      term5(it)=  c1*exp(r1*(tout(it)-1850));
    end
%
% Numerical output
    fprintf('\n          t                        ');
    fprintf('\n      term1     term2     temp3     term4');
    fprintf('\n      term5     dyla/dt     yla        ');
    for(it=1:nout)
      fprintf('\n\n t = %5.0f',tout(it));
      fprintf('\n%10.4f%10.4f%10.4f%10.4f',...
              term1(it),term2(it),term3(it),term4(it));
```

```
    fprintf('\n%10.4f%10.4f%10.4f',...
        term5(it),dyla(it),yla(it));
  end
%
% Graphical output
  figure(8);
  subplot(3,2,1);
  plot(tout,term1,'-');axis([1850 2100 -0.1 0.1]);
  xlabel('t (year)');ylabel('term1');
  title('1/5*(yua(it)-yla(it))');
  subplot(3,2,2);
  plot(tout,term2,'-');axis([1850 2100 -0.1 0.1]);
  xlabel('t (year)');ylabel('term2');
  title('1/0.75*(ysb(it)-yla(it))');
  subplot(3,2,3)
  plot(tout,term3,'-');axis([1850 2100 -0.1 0.1]);
  xlabel('t (year)');ylabel('term3');
  title('1/150*(ylb(it)-yla(it))');
  subplot(3,2,4)
  plot(tout,term4,'-');axis([1850 2100 -0.1 0.1]);
  xlabel('t (year)');ylabel('term4');
  title('1/30*(yul(it)-yla(it))');
  subplot(3,2,5)
  plot(tout,term5,'-');axis([1850 2100 -0.1 0.2]);
  xlabel('t (year)');ylabel('term5');
  title('c1*exp(r1*(tout(it)-1850))');
  subplot(3,2,6)
  plot(tout,dyla,'-');axis([1850 2100 -0.1 0.1]);
  xlabel('t (year)');ylabel('dyla(t)/dt');
  title('dyla(t)/dt');
```

We can note the following details about these main programs.

- The coding is the same as in Listings 1a, 1b up to and including the statement co2s=0.005. Note in particular the selection ncase=1 (ncase=1,2,3,4 can be used).

R:

```
#
# CO2 sensitivity (C^o/ppm CO2)
  co2s=0.005;
#
# Compute and plot derivatives, RHS terms
```

Matlab:

```
%
% CO2 sensitivity (C^o/ppm CO2)
  co2s=0.005;
%
% Compute and plot derivatives, RHS terms
```

The ODE routines model_1.R, model_1.m (Listings 3a and 3b), the CO_2 emission rate routines co2_rate.R, co2_rate.m (Listings 4a and 4b), and the spline routines spline.R, spline.m (Listings 5a and 5b) are unchanged.

- The dependent variable of Eqs. (3.3) to (3.9) are placed in arrays for subsequent calculations and plotting (with a for in it for nout=26 output points in t).

R:

```
#
# Model dependent variables
  yla=rep(0,nout);yua=rep(0,nout);ysb=rep(0,nout);
  ylb=rep(0,nout);yul=rep(0,nout);ydl=rep(0,nout);
  ymb=rep(0,nout);
  for(it in 1:nout){
    yla[it]=out[it,2];yua[it]=out[it,3];
    ysb[it]=out[it,4];ylb[it]=out[it,5];
    yul[it]=out[it,6];ydl[it]=out[it,7];
    ymb[it]=out[it,8];
  }
```

For R, arrays are first declared (preallocated) using the rep utility. The 2D array out comes from the numerical integration of ode (see Listing 1e). Note the offset of 1 in the second subscript since the subscript 1 pertains to t (a feature of the integrator ode).

Matlab:

```
%
% Model dependent variables
  for(it=1:nout)
    yla(it)=y(it,1);yua(it)=y(it,2);
    ysb(it)=y(it,3);ylb(it)=y(it,4);
```

```
yul(it)=y(it,5);ydl(it)=y(it,6);
ymb(it)=y(it,7);
end
```

Declaration of the arrays is not required in Matlab.

- The derivatives of Eqs. (3.3) to (3.9) are computed and placed in arrays, e.g., from Eq. (3.4a), $dy_{la}/dt =$ dyla.

R:

```
#
# Model dependent variable derivatives
dyla=rep(0,nout);dyua=rep(0,nout);dysb=rep(0,nout);
dylb=rep(0,nout);dyul=rep(0,nout);dydl=rep(0,nout);
dymb=rep(0,nout);
for(it in 1:nout){
  CO2_rate(tout[it],ncase);
  dyla[it]=    1/5*(yua[it]-yla[it])+
               1/0.75*(ysb[it]-yla[it])+
               1/150*(ylb[it]-yla[it])+
               1/30*(yul[it]-yla[it])+
               c1*exp(r1*(tout[it]-1850));
  dyua[it]=    1/3*(yla[it]-yua[it]);
  dysb[it]=  1/0.75*(yla[it]-ysb[it]);
  dylb[it]=  1/150*(yla[it]-ylb[it]);
  dyul[it]=    1/80*(yla[it]-yul[it])+
               1/200*(ydl[it]-yul[it])+
               1/5*(ymb[it]-yul[it]);
  dydl[it]=  1/1500*(yul[it]-ydl[it]);
  dymb[it]=    1/10*(yul[it]-ymb[it]);
}
```

Again, declaration of arrays (with **rep**) is required in R. The CO_2 emissions rate **co2_rate.R** of Listing 4a is called first. Then the coding of the ODEs in **model_1.R** of Listing 3a is used to compute the derivatives of Eqs. (3.3) to (3.9). The source term of Eq. (3.10) is used for the programming of Eq. (3.4a). The final } concludes the **for**.

Matlab:

```
%
% Model dependent variable derivatives
  for(it=1:nout)
    [c1,r1]=CO2_rate(tout(it));
    dyla(it)=      1/5*(yua(it)-yla(it))+...
                   1/0.75*(ysb(it)-yla(it))+...
                   1/150*(ylb(it)-yla(it))+...
                   1/30*(yul(it)-yla(it))+...
                   c1*exp(r1*(tout(it)-1850));
    dyua(it)=      1/3*(yla(it)-yua(it));
    dysb(it)=  1/0.75*(yla(it)-ysb(it));
    dylb(it)=   1/150*(yla(it)-ylb(it));
    dyul(it)=    1/80*(yla(it)-yul(it))+...
                 1/200*(ydl(it)-yul(it))+...
                   1/5*(ymb(it)-yul(it));
    dydl(it)=  1/1500*(yul(it)-ydl(it));
    dymb(it)=    1/10*(yul(it)-ymb(it));
  end
```

Again, a `for` is used to step through the `nout=26` output points
in t.

- Two plots on the same page are produced for each dependent variable of Eqs. (3.3) to (3.9) and the corresponding derivative.

R:

```
#
# Plot numerical solutions, derivatives
#
# yla, dyla/dt
  par(mfrow=c(1,2))
  plot(tout,yla,xlab="t (year)",ylab="yla(t)",
     xlim=c(1850,2100),main="yla",type="l",lwd=2);
  plot(tout,dyla,xlab="t (year)",ylab="dyla(t)/dt",
     xlim=c(1850,2100),main="dyla/dt",type="l",lwd=2);
```

Note the use of `par(mfrow=c(1,2))` for a $1(row) \times 2(column)$ array of plots. The two calls to `plot` produce the two plots, e.g., for y_{la} and dy_{la}/dt.

Matlab:

```
%
% Plot numerical solutions, derivatives
%
% yla, dyla/dt
  figure(1);
  subplot(1,2,1);
  plot(tout,yla,'-');
  xlabel('t (year)');ylabel('yla(t)');
  title('yla');axis tight;
  subplot(1,2,2);
  plot(tout,dyla,'-');
  xlabel('t (year)');ylabel('dyla(t)/dt');
  title('dyla/dt');axis tight;
         .                    .
         .                    .
         .                    .
```

The 1×2 array of plots is produced with the Matlab `subplot` utility e.g., `subplot(1,2,1)` declares a 1×2 array, plot no. 1. Similary for the second (derivative) plot, `subplot(1,2,2)`.

The first plot of $y_{la}, dy_{la}/dt$ vs t (Figs. 10a and 10b), and the seventh plot of $y_{mb}, dy_{mb}/dt$ versus t (Figs. 11a and 11b) follow.

We observe in each of these plots that the derivative in t remains positive throughout $1850 \leq t \leq 2100$ and the corresponding dependent variable increases monotonically throughout the full interval in t. Also, the derivative dy_{la}/dt has a departure from a smooth curve near $t = 1850$, but the numerical integration of this derivative gives a smooth solution curve for y_{la}. The non-smooth dy_{la}/dt is not unusual since at the beginning of the solution, there can be a departure of the solution from the ICs to values given by the ODEs. Also, this example demonstrates the smoothing effect of numerical integration.

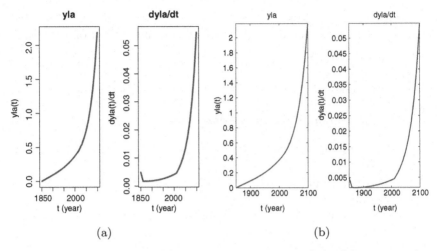

(a)   (b)

Figure 10: (a) $y_{la}, dy_{la}/dt$ vs t from Listing 1e (R). (b) $y_{la}, dy_{la}/dt$ vs t from Listing 1f (Matlab)

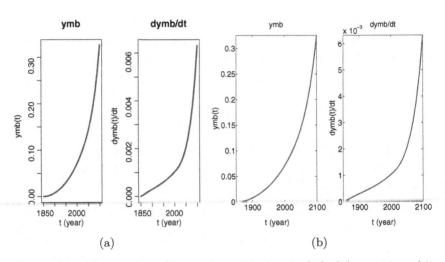

(a)   (b)

Figure 11: (a) $y_{mb}, dy_{mb}/dt$ vs t from Listing 1e (R). (b) $y_{mb}, dy_{mb}/dt$ vs t from Listing 1f (Matlab)

- The individual RHS terms of Eq. (3.4a) are placed in arrays. This code follows directly from the preceding coding for Eq. (3.4a) (and the code in Listings 3a and 3b).

R:

```
#
# RHS terms
  term1=rep(0,nout);term2=rep(0,nout);term3=rep(0,nout);
  term4=rep(0,nout);term5=rep(0,nout);
  for(it in 1:nout){
    CO2_rate(tout[it],ncase);
    term1[it]=    1/5*(yua[it]-yla[it]);
    term2[it]= 1/0.75*(ysb[it]-yla[it]);
    term3[it]= 1/150*(ylb[it]-yla[it]);
    term4[it]=  1/30*(yul[it]-yla[it]);
    term5[it]=  c1*exp(r1*(tout[it]-1850));
  }
```

Matlab:

```
%
% RHS terms
  for(it=1:nout)
    [c1,r1]=CO2_rate(tout(it));
    term1(it)=    1/5*(yua(it)-yla(it));
    term2(it)= 1/0.75*(ysb(it)-yla(it));
    term3(it)= 1/150*(ylb(it)-yla(it));
    term4(it)=  1/30*(yul(it)-yla(it));
    term5(it)=  c1*exp(r1*(tout(it)-1850));
  end
```

- Numerical output includes the RHS terms, `term1` to `term5`, dy_{la}/dt and y_{la}.

R:

```
#
# Numerical output
  cat(sprintf(
    "\n        t                              "));
  cat(sprintf(
    "\n     term1      term2      temp3      term4"));
  cat(sprintf(
    "\n      term5    dyla/dt        yla       "));
  for(it in 1:nout){
    cat(sprintf("\n\n t = %5.0f",tout[it]));
    cat(sprintf("\n%10.4f%10.4f%10.4f%10.4f",
```

```
        term1[it],term2[it],term3[it],term4[it]));
    cat(sprintf("\n%10.4f%10.4f%10.4f",
        term5[it],dyla[it],yla[it]));
}
```

Matlab:

```
%
% Numerical output
    fprintf('\n        t                        ');
    fprintf('\n      term1      term2      temp3      term4');
    fprintf('\n      term5      dyla/dt    yla              ');
    for(it=1:nout)
      fprintf('\n\n t = %5.0f',tout(it));
      fprintf('\n%10.4f%10.4f%10.4f%10.4f',
          term1(it),term2(it),term3(it),term4(it));
      fprintf('\n%10.4f%10.4f%10.4f',
          term5(it),dyla(it),yla(it));
    end
```

- Plotting of the five RHS terms of Eq. (3.4a) and dy_{la}/dt (as single page 3×2 matrices) concludes the main programs.

R:

```
#
# Graphical output
    par(mfrow=c(3,2))
    plot(tout,term1,xlab="t (year)",ylab="term1",
        xlim=c(1850,2100),ylim=c(-0.1,0.1),
        main="1/5*(yua(t)-yla(t))",type="l",lwd=2);
    plot(tout,term2,xlab="t (year)",ylab="term2",
        xlim=c(1850,2100),ylim=c(-0.1,0.1),
        main="1/0.75*(ysb(t)-yla(t))",type="l",lwd=2);
    plot(tout,term3,xlab="t (year)",ylab="term3",
        xlim=c(1850,2100),ylim=c(-0.1,0.1),
        main="1/150*(ylb[it]-yla[it])",type="l",lwd=2);
    plot(tout,term4,xlab="t (year)",ylab="term4",
        xlim=c(1850,2100),ylim=c(-0.1,0.1),
        main="1/30*(yul(t)-yla(t))",type="l",lwd=2);
    plot(tout,term5,xlab="t (year)",ylab="term5",
        xlim=c(1850,2100),ylim=c(-0.1,0.2),
        main="1.1*exp(r1*(t-1850))",type="l",lwd=2);
```

```
plot(tout,dyla,xlab="t (year)",ylab="dla(t)/dt",
    xlim=c(1850,2100),ylim=c(-0.1,0.1),
    main="dla(t)/dt",type="l",lwd=2);
```

A 3×2 array of six plots is specified with `par(mfrow=c(3,2))`. The sixth plot is dy_{la}/dt vs t to indicate the overall direction of this derivative with t. The contributions of the RHS terms to this derivative are indicated in the preceding five plots. In this way, the relative contribution of each term is clear. Note that to facilitate this type of comparison, the vertical scale is set to `ylim=c(-0.1,0.1)` except for the `term5` graph which required `ylim=c(-0.1,0.2)`. Thus, `term5` is the largest term indicating the dominant effect of the source term of Eq. (3.10) as explained in the subsequent numerical output.

Matlab:

```
figure(8);
subplot(3,2,1);
plot(tout,term1,'-');axis([1850 2100 -0.1 0.1]);
xlabel('t (year)');ylabel('term1');
title('1/5*(yua(it)-yla(it))');
subplot(3,2,2);
plot(tout,term2,'-');axis([1850 2100 -0.1 0.1]);
xlabel('t (year)');ylabel('term2');
title('1/0.75*(ysb(it)-yla(it))');
subplot(3,2,3)
plot(tout,term3,'-');axis([1850 2100 -0.1 0.1]);
xlabel('t (year)');ylabel('term3');
title('1/150*(ylb(it)-yla(it))');
subplot(3,2,4)
plot(tout,term4,'-');axis([1850 2100 -0.1 0.1]);
xlabel('t (year)');ylabel('term4');
title('1/30*(yul(it)-yla(it))');
subplot(3,2,5)
plot(tout,term5,'-');axis([1850 2100 -0.1 0.2]);
xlabel('t (year)');ylabel('term5');
title('c1*exp(r1*(tout(it)-1850))');
subplot(3,2,6)
plot(tout,dyla,'-');axis([1850 2100 -0.1 0.1]);
xlabel('t (year)');ylabel('dyla(t)/dt');
title('dyla(t)/dt');
```

Again, a 3×2 array of plots is specified. The first of the six plots follows from subplot(3,2,1); and the sixth plot for the derivative of Eq. (3.4a), dy_{la}/dt, from subplot(3,2,6). Note again the common scaling axis([1850 2100 -0.1 0.1]) except for the fifth plot.

A discussion of the numerical and graphical output from the preceding code is next.

Abbreviated numerical output from Listings 1e and 1f is in Tables 7a and 7b.

Table 7a: Abbreviated numerical output from the main program of Listing 1e

R:

```
ncase =    1
```

t	yla	yul	ppm	Tc	eps	pH	r1
1850	0.0000	0.0000	280.0	1.745	8.800	8.222	0.0100

```
             .                           .
             .                           .
             .                           .
    Output for t = 1860 to 2090 removed; see Tables 3a, 6a
             .                           .
             .                           .
             .                           .
```

| 2100 | 2.1869 | 0.3886 | 892.3 | 1.940 | 14.422 | 7.795 | 0.0150 |

```
Atmosphere ppm(2013) =    400.2

GtC Lower Atmosphere(2013) =   725.3

GtC Upper Atmosphere(2013) =   126.7

GtC Long-lived(2013)       = 2498.9

GtC Short-lived(2013)      =  156.8

GtC Upper Layer(2013)      =  987.2

GtC Deep Layer(2013)       = 37227.5
```

```
GtC Marine Biota(2013)      =     3.3

ncall =   363

         t
    term1      term2      temp3      term4
    term5      dyla/dt     yla

t =  1850
    0.0000     0.0000     0.0000     0.0000
    0.0049     0.0049     0.0000

t =  1860
   -0.0011    -0.0018    -0.0001    -0.0006
    0.0054     0.0018     0.0200
                 .          .
                 .          .
                 .          .
Output for t = 1870 to t = 1990 removed
                 .          .
                 .          .
                 .          .
t =  2000
   -0.0024    -0.0041    -0.0017    -0.0096
    0.0220     0.0042     0.3693

t =  2010
   -0.0027    -0.0046    -0.0019    -0.0107
    0.0244     0.0046     0.4131

t =  2020
   -0.0034    -0.0060    -0.0021    -0.0120
    0.0296     0.0061     0.4664
                 .          .
                 .          .
                 .          .
Output for t = 2030 to t = 2080 removed
                 .          .
                 .          .
                 .          .
t =  2090
   -0.0224    -0.0397    -0.0084    -0.0465
```

```
    0.1576      0.0406      1.7142

t =   2100
   -0.0301     -0.0535     -0.0109     -0.0599
    0.2092      0.0547      2.1867
```

Table 7b: Abbreviated numerical output from the main program of Listing 1f

Matlab:

```
ncase =    1

   t      yla      yul      ppm      Tc      eps      pH      r1
1850   0.0000   0.0000    280.0    1.745   8.800   8.222   0.0100
              .                             .
              .                             .
              .                             .
    Output for t = 1860 to 2090 removed; see Tables 3b, 6b
              .                             .
              '                             .
              .                             .
2100   2.1869   0.3886    892.3    1.940  14.422   7.795   0.0150

Atmosphere ppm(2013) =    400.2

GtC Lower Atmosphere(2013) =    725.3

GtC Upper Atmosphere(2013) =    126.7

GtC Long-lived(2013)           -  2498.9

GtC Short-lived(2013)      =    156.8

GtC Upper Layer(2013)      =    987.2

GtC Deep Layer(2013)       = 37227.5

GtC Marine Biota(2013)     =      3.3

ncall =    153
```

```
        t
    term1      term2      temp3      term4
    term5      dyla/dt      yla


t =  1850
    0.0000     0.0000     0.0000     0.0000
    0.0049     0.0049     0.0000


t =  1860
   -0.0011    -0.0018    -0.0001    -0.0006
    0.0054     0.0018     0.0200
                 .          .
                 .          .
                 .          .
Output for t = 1870 to t = 1990 removed
                 .          .
                 .          .
                 .          .
t =  2000
   -0.0024    -0.0041    -0.0017    -0.0096
    0.0220     0.0042     0.3693


t =  2010
   -0.0027    -0.0046    -0.0019    -0.0107
    0.0244     0.0046     0.4131


t =  2020
   -0.0034    -0.0060    -0.0021    -0.0120
    0.0296     0.0061     0.4667
                 .          .
                 .          .
                 .          .
Output for t = 2030 to t = 2080 removed
                 .          .
                 .          .
                 .          .
t =  2090
   -0.0224    -0.0397    -0.0084    -0.0465
    0.1576     0.0406     1.7143


t =  2100
   -0.0301    -0.0535    -0.0109    -0.0599
    0.2092     0.0547     2.1869
```

We can note the following details of this output.

- At the beginning of the solutions, $t = 1850$,

  R:

t			
term1	term2	temp3	term4
term5	dyla/dt	yla	

t = 1850			
0.0000	0.0000	0.0000	0.0000
0.0049	0.0049	0.0000	

  Matlab:

t			
term1	term2	temp3	term4
term5	dyla/dt	yla	

t = 1850			
0.0000	0.0000	0.0000	0.0000
0.0049	0.0049	0.0000	

  The zero initial values in the first row follow from the zero ICs for the ODEs, Eqs. (3.3) to (3.9). That is, from the RHS terms of Listing 1e,

  ```
  term1[it]=    1/5*(yua[it]-yla[it]);
  term2[it]= 1/0.75*(ysb[it]-yla[it]);
  term3[it]=  1/150*(ylb[it]-yla[it]);
  torm4[it]-   1/30*(yul[it]-yla[it]);
  ```

  all of the dimensionless concentrations are zero at $t = 1850$.

- term5 = 0.0049 is the numerical value of the source term from Eq. (3.10), that is, from Listing 1e

  ```
  term5[it]=  c1*exp(r1*(tout(it)-1850))=c1*exp(0)=c1=0.0049
  ```

In summary, for Eq. (3.4a) at $t = 1850$ (see also Fig. 1),

reservoir exchange, emissions rate, derivative, concentration	R coding and value
lower atmosphere upper atmosphere	`term1[it]=1/5*(yua[it]-yla[it])=0`
lower atmosphere short-lived biota	`term2[it]=1/0.75*(ysb[it]-yla[it])=0`
lower atmosphere long-lived biota	`term3[it]=1/150*(ylb[it]-yla[it])=0`
lower atmosphere ocean upper layer	`term4[it]=1/30*(yul[it]-yla[it])=0`
lower atmosphere emissions rate	`term5[it]=c1*exp(r1*(tout(it)-1850))=` `0.0049`
dy_{la}/dt	`term1[it]+term2[it]+term3[it]+` `term4[it]+term5[it]=0.0049`
lower atmosphere concentration	y_{la}=`0.0000`

- Similarly, at the intermediate point $t = 2010$ (from Tables 7a and 7b),

  R:

  ```
        t
    term1     term2      temp3      term4
    term5    dyla/dt      yla
  ```

```
t =   2010
   -0.0027   -0.0046   -0.0019   -0.0107
    0.0244    0.0046    0.4131
```

Matlab:

```
        t
   term1      term2      temp3      term4
   term5      dyla/dt     yla

t =   2010
   -0.0027   -0.0046   -0.0019   -0.0107
    0.0244    0.0046    0.4131
```

In summary, for Eq. (3.4a) at $t = 2010$,

reservoir exchange,   R coding and value
emissions rate,
derivative,
concentration

lower atmosphere   `term1[it]=1/5*(yua[it]-yla[it])=`
upper atmosphere   `-0.0027`

lower atmosphere   `term2[it]=1/0.75*(ysb[it]-yla[it])=`
short-lived biota   `-0.0046`

lower atmosphere   `term3[it]=1/150*(ylb[it]-yla[it])=`
long-lived biota   `-0.0019`

lower atmosphere   `term4[it]=1/30*(yul[it]-yla[it])=`
ocean upper layer   `-0.0107`

lower atmosphere   `term5[it]=c1*exp(r1*(tout(it)-1850))=`
emissions rate   `0.0244`

dy_{la}/dt `term1[it]+term2[it]+term3[it]+`
                             `term4[it]+term5[it]=0.0046`

lower atmosphere y_{la}=`0.4131`
concentration

While `term1` to `term4` are negative they do not have a large enough absolute sum to equal the positive value of `term5` = `0.0244` so the derivative dy_{la}/dt = `0.0046` remains positive and y_{la} continues to increase with $y_{la}(2010)$ = `0.4131`.

The same general conclusions apply at 2100, that is, dy_{la}/dt remains positive and $y_{la}(2100)$ = `2.1867` (from Table 7a), so that a 218% increase in the CO_2 concentration to 892.3 ppm occurs from 280 at 1850 (from Tables 3a and 3b).

The grapical output is next in Figs. 12a and 12b which reflects the numerical output in Tables 7a and 7b.

In general, although `term1` to `term4` are negative reflecting a movement of CO_2 from the lower atmosphere to four other reservoirs, the combined (summed) effect of these four terms is not enough to offset the emissions rate of `term5` so that y_{la} continues to increase throughout $1850 \leq t \leq 2100$ and dy_{la}/dt remains positive as in Figs. 10a and 10b.

As a concluding example, we repeat the preceding analysis for `ncase=4` in Listings 1e and 1f and observe that up to approximately 2050, `term1` to `term4` have an absolute sum less than `term5` so CO_2 continues to accumulate in the lower atmosphere. However, beyond $t = 2050$ `term1` to `term4` have an absolute sum greater than `term5` so that more CO_2 leaves the lower atmosphere than enters it through the emissions of `term5`, and $dy_{la}/dt < 0$ (a sign change in this derivative). Thus, CO_2 begins to leave the lower atmosphere (by moving to the other six reservoirs). This maximum in y_{la} is clear in Figs. 12a, 12b for `ncase=4`.

Also, the following brief output from the main programs of Listings 1e and 1f (with `ncase=4`) confirms the change in the sign of dy_{ya}/dt.

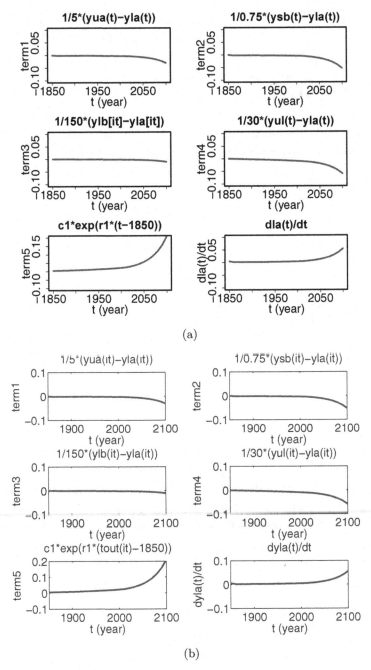

(a)

(b)

Figure 12: (a) $term1, \ldots, term5, dy_{la}/dt$ vs t from Listing 1e (R).
(b) $term1, \ldots, term5, dy_{la}/dt$ vs t from Listing 1f (Matlab)

R:

```
t =   2030
    -0.0015    -0.0022    -0.0020    -0.0120
     0.0200     0.0022     0.4795

t =   2040
    -0.0009    -0.0013    -0.0020    -0.0121
     0.0175     0.0012     0.4962

t =   2050
    -0.0003    -0.0004    -0.0019    -0.0120
     0.0149     0.0003     0.5037

t =   2060
     0.0001     0.0003    -0.0018    -0.0115
     0.0125    -0.0004     0.5032

t =   2070
     0.0005     0.0009    -0.0016    -0.0110
     0.0102    -0.0010     0.4964

t =   2080
     0.0008     0.0014    -0.0015    -0.0102
     0.0082    -0.0014     0.4846
```

Matlab:

```
t =   2030
    -0.0015    -0.0023    -0.0020    -0.0120
     0.0200     0.0022     0.4789

t =   2040
    -0.0009    -0.0013    -0.0020    -0.0121
     0.0175     0.0012     0.4957

t =   2050
    -0.0004    -0.0004    -0.0019    -0.0119
     0.0149     0.0003     0.5032

t =   2060
     0.0001     0.0003    -0.0018    -0.0115
     0.0125    -0.0004     0.5028
```

```
t =   2070
   0.0005     0.0009    -0.0016    -0.0109
   0.0102    -0.0010     0.4961


t =   2080
   0.0008     0.0013    -0.0015    -0.0102
   0.0082    -0.0014     0.4843
```

Note in particular the change in the sign of dy_{la}/dt (for R, $dy_{la}(2030)/dt = 0.0022$ and $dy_{la}(2080)/dt = -0.0014$) and the maximum in y_{la} near $t = 2050$ (for R, $y_{la}(2050) = 0.5037$). The change in the derivative for ncase=4 results from the reduction of the exponential rate constant r_1 of Eq. (3.10) as calculated in CO2_rate.R and CO2_rate.m of Listings 4a and 4b. The important conclusion for ncase=4 is that the atmospheric CO_2 concentration can be reduced if the emissions rate is reduced sufficiently and some of the atmospheric CO_2 moves to the other four reservoirs. Whether such a reduction in the CO_2 emissions rate can be achieved remains to be seen.

The intent of the preceding analysis is to demonstrate how the individual RHS terms in an ODE system can be computed and displayed to achieve enhanced insight into the dynamic characteristics of the model as reflected in the LHS derivatives. In other words, the model derivatives given by the ODEs tell the model what to do (the response of the dependent variables). This type of study can then be extended by, for example, varying the model parameters such as the $\theta's$ in Eqs. (3.3) to (3.9) (as programmed in the main programs of Listings 1e and 1f and the ODE routines of Listings 3a and 3b).

APPENDIX B: ADDITIONAL ANALYSIS OF THE MODEL ODEs

Some additional features of the ODE model of Eqs. (3.3) to (3.10) are considered in this appendix.

B1. Some Features of the ODE Model

The ordinary differential equations (ODEs) in the introductory global CO_2 model are based on *dimensionless dependent variables* of the form

$$y_i(t) = \frac{c_i(t) - c_i(t = t_0)}{c_i(t = t_0)}, \qquad (B1)$$

where

Variable	Definition
$y_i(t)$	dimensionless dependent variable for reservoir i (e.g., dimensionless carbon concentration)
$c_i(t)$	dimensional dependent variable for reservoir i (e.g., dimensional carbon concentration)
t	time (calendar year)
t_0	initial value of t (typically, 1850)

The model ODEs based on the dimensionless dependent variable of Eq. (B1) are of the form

$$\frac{dy_i}{dt} = \frac{1}{\theta_j^i}(y_j - y_i), \qquad (B2)$$

173

where θ_j^i is the mixing time or mean residence time of reservoir i for transfer from reservoir j, and y_j is the dependent variable for reservoir j.

As noted previously, the use of dimensionless variables of the form of Eq. (B1) has significant advantages:

- All of the model dependent variables have a common scale, the fractional change from the initial value, which facilitates comparison of the variables such as in composite plots.
- All of the model dependent variables start from homogeneous (zero) initial conditions.
- Carbon fluxes between the reservoirs are expressed in terms of the mean residence times (θ for each reservoir) so that fluxes in terms of carbon flows are not required.
- Carbon inventories and fluxes that are used to estimate θ_j^i in Eq. (B2) (considered next) do not require conversion of units.

However, the ODEs of the form of Eq. (B2) can also be expressed in dimensional form which can provide a clearer physical interpretation. Thus, combining Eqs. (B1) and (B2) gives

$$\frac{d\left[\dfrac{c_i - c_i(t_0)}{c_i(t_0)}\right]}{dt} = \frac{1}{\theta_j^i}\left[\frac{c_j - c_j(t_0)}{c_j(t_0)} - \frac{c_i - c_i(t_0)]}{c_i(t_0)}\right],$$

$$\frac{d\left[\dfrac{c_i}{c_i(t_0)} - 1\right]}{dt} = \frac{1}{\theta_j^i}\left[\left(\frac{c_j}{c_j(t_0)} - 1\right) - \left(\frac{c_i}{c_i(t_0)} - 1\right)\right],$$

$$\frac{d\left[\dfrac{c_i}{c_i(t_0)}\right]}{dt} = \frac{1}{\theta_j^i}\left[\left(\frac{c_j}{c_j(t_0)}\right) - \left(\frac{c_i}{c_i(t_0)}\right)\right],$$

$$\frac{dc_i}{dt} = \frac{1}{\theta_j^i}\left[\frac{c_i(t_0)}{c_j(t_0)}c_j - c_i\right],$$

$$\frac{dc_i}{dt} = \frac{Q_i^j}{V_i}\frac{c_i(t_0)}{c_j(t_0)}c_j - \frac{Q_i^j}{V_i}c_i,$$

where

Variable	Definition
Q_i^j	volumetric flow from reservoir i to reservoir j
V_i	volume of reservoir i

and $\theta_j^i = V_i/Q_i^j$.

If we assume that initially (at $t = t_0$) there is no net CO_2 flux between reservoirs i and j (a net flux will develop only after anthropogenic CO_2 is added to the lower atmosphere)

$$Q_j^i c_j(t_0) = Q_i^j c_i(t_0) \tag{B3}$$

or $Q_j^i = \dfrac{c_i(t_0)}{c_j(t_0)} Q_i^j$ and substitution in the preceding ODE gives

$$\frac{dc_i}{dt} = \frac{Q_j^i}{V_i} c_j - \frac{Q_i^j}{V_i} c_i$$

or

$$V_i \frac{dc_i}{dt} = Q_j^i c_j - Q_i^j c_i. \tag{B4}$$

Equation (B4) is the conventional material balance for a well mixed reservoir. In other words, the dimensionless ODE, Eq. (B2), can be considered as a material balance subject to the condition of Eq. (B3). Similar reasoning applies to the case of multiple CO_2 streams into and out of a reservoir as, for example, the lower atmosphere and ocean upper layer.

B2. Linear Algebra

The ODE system of Eqs. (3.3) to (3.9) with Q_c from Eq. (3.10) can be classified as a 7×7 (seven ODEs in seven unknowns), linear, constant coefficient, inhomogeneous (from Q_c) ODE system that has a solution $y_1(t), \ldots, y_7(t)$. Thus, this ODE system has the characteristics and features that can be studied by a few basic concepts from linear algebra. The intent of the following discussion is to explore this (linear algebra) approach to the analysis of Eqs. (3.3) to (3.10). Equations (3.3) to (3.10) can be expressed in matrix for as

$$\frac{d\mathbf{y}}{dt} = \mathbf{cy} + \mathbf{q}, \qquad (B5.1)$$

where

$$\mathbf{y} = \begin{bmatrix} y_{la} & y_{ua} & y_{sb} & y_{lb} & y_{ul} & y_{dl} & y_{mb} \end{bmatrix}^T, \qquad (B5.2)$$

$$dy/dt = \begin{bmatrix} dy_{la}/dt & dy_{ua}/dt & dy_{sb}/dt & dy_{lb}/dt & dy_{ul}/dt \\ dy_{dl}/dt & dy_{mb}/dt \end{bmatrix}^T, \qquad (B5.3)$$

$$\mathbf{q} = \begin{bmatrix} Q_c & 0 & 0 & 0 & 0 & 0 & 0 \end{bmatrix}^T, \qquad (B5.4)$$

Q_c from Eq. (3.10) is used in Eq. (B5.4).

$$
\mathbf{c} = \begin{bmatrix}
(-1/5 - 1/0.75 & 1/5 & 1/0.75 & 1/150 & 1/30 & 0 & 0 \\
-1/150 - 1/30) & & & & & & \\
1/3 & -1/3 & 0 & 0 & 0 & 0 & 0 \\
1/0.75 & 0 & 1/0.75 & 0 & 0 & 0 & 0 \\
1/150 & 0 & 0 & -1/150 & 0 & 0 & 0 \\
1/80 & 0 & 0 & 0 & (-1/80 - 1/200 & 1/200 & 1/5 \\
 & & & & -1/5) & & \\
0 & 0 & 0 & 0 & 1/1500 & -1/1500 & 0 \\
0 & 0 & 0 & 0 & 1/10 & 0 & -1/10
\end{bmatrix}
$$

$$(B5.5)$$

The 7×7 matrix $\mathbf{c}$ of Eq. (B5.5) follows from the coding of the seven ODEs in `model_1.R` of Listing 3a and in `model_1.m` of Listing 3b.

```
CO2_rate(t,ncase);
dyla=     1/5*(yua-yla)+
          1/0.75*(ysb-yla)+
          1/150*(ylb-yla)+
          1/30*(yul-yla)|
```

```
        c1*exp(r1*(t-1850)));
dyua=   1/3*(yla-yua);
dysb=   1/0.75*(yla-ysb);

dylb=   1/150*(yla-ylb);
dyul=   1/80*(yla-yul)+
        1/200*(ydl-yul)+
        1/5*(ymb-yul);
dydl=   1/1500*(yul-ydl);
dymb=   1/10*(yul-ymb);
```

Note that the columns of c (Eq. (B5.5)) correspond to the elements of y of Eq. (B5.2) and the rows correspond to the elements of dy/dt in Eq. (B5.3).

The analytical solution to Eqs. (B5) is

$$y(t) = y_h(t) + y_p(t), \tag{B6.1}$$

where $y_h(t)$ is the homogeneous solution to Eq. (B5.1) (with $q = 0$) and $y_p(t)$ is a particular solution (for a particular q, e.g., from Eq. (B5.4)). $y_h(t)$ is a linear combination of exponential functions in t

$$y_h(t) = c_1 e^{\lambda_1 t} + c_2 e^{\lambda_2 t} + c_3 e^{\lambda_3 t} + c_4 e^{\lambda_4 t} + c_5 e^{\lambda_5 t} + c_6 e^{\lambda_6 t} + c_7 e^{\lambda_7 t}$$

$$= \sum_{i=1}^{7} c_i e^{\lambda_i} \tag{B6.2}$$

λ_i, $i = 1, \ldots, 7$ is a vector of seven eigenvalues of c of Eq. (B5.5), that is, the roots of the seventh order polynomial (characteristic equation)

$$det(c - \lambda I) = 0, \tag{B6.3}$$

where det denotes a determinant and I is the 7×7 identity matrix. c_i (in Eq. (B6.2)) is the eigenvector for eigenvalue λ_i.

For the solution to Eqs. (3.3) to (3.10) to be stable and nonoscillatory (as we would expect and as reflected in the solutions for $y_{la}(t)$ of Figs. 6a and 6b), λ_i must be real and nonpositive. This can be explored by numerically computing λ_i and c_i in Eq. (B6.2). Eigenvalue/eigenfunction utilities in R and Matlab are available for this standard, linear algebra calculation.

We conclude by illustrating the preceding ideas applied to the 1×1 system of Eq. (B2) where y_i is the dependent variable of this ODE and y_j is an input or forcing function. Equation (B2) slightly rearranged is

$$\frac{dy_i}{dt} = -\frac{1}{\theta_j^i} y_i + \frac{1}{\theta_j^i} y_j. \tag{B7.1}$$

If we take $y_j = 1$ (a unit input or forcing function) and drop the i, j designation,

$$\frac{dy}{dt} = -\frac{1}{\theta} y + \frac{1}{\theta}. \tag{B7.2}$$

Equation (B7.2) is a linear, constant coefficient, inhomogeneous ODE with the general solution of the form of Eqs. (B6.1) and (B6.2). Specifically,

$$y_h(t) = ce^{-(1/\theta)t}; \quad y_p = 1, \tag{B7.3}$$

where c is a constant (one-element eigenvector) determined from the homogeneous (zero) initial condition (IC) $y(t = 0) = 0$, that is, $c = -1$.

The solution to Eq. (B7.3) is then

$$y(t) = 1 - e^{-(1/\theta)t}. \tag{B7.4}$$

A comparison of Eqs. (B6.2) and (B7.4) (for one first-order ODE) indicates: (1) $\lambda_1 = \lambda = -1/\theta$, (2) $\mathbf{c_1} = -\mathbf{1}$. In particular, the solution, Eq. (B7.4), (1) is stable (since $\theta > 0$), (2) has the homogeneous IC $y(t = 0) = 0$ (3) has a time scale defined by θ and (4) approaches the limiting value $y(t \to \infty) = 1$ (as determined by the input or forcing function $y_j = 1$). (1) also follows from Eq. (B6.3)

$$det(-1/\theta - \lambda) = 0$$

or $\lambda = -1/\theta$.

The intent of this discussion is to illustrate the role of ODE eigenvalues and eigenvectors in determining the features of a system of linear, constant coefficient, nonhomogeneous ODEs, and in particular Eqs. (3.3) to (3.10). The 7×7 matrix formulation of Eqs. (B5) follows the approach in [2].

B3. Estimation of Residence Times

The preceding discussion of Section B2 indicates that the θ's in Eqs. (3.3) to (3.9) define the time scales of the various carbon transfers between the seven reservoirs. The specific numerical values of the θ's are programmed in functions model_1.R and model_1.m of Listings 3a and 3b. The numerical estimation of each specific θ_j^i (termed a residence time, time constant, turnover time) in Eq. (B2) is now considered. For example, the programming of Eq. (3.4a) includes four values of θ_j^i

```
dyla=     1/5*(yua-yla)+
          1/0.75*(ysb-yla)+
          1/150*(ylb-yla)+
          1/30*(yul-yla)+
          c1*exp(r1*(t-1850));
```

Each θ_j^i is interpreted as a time scale for the corresponding carbon transfer, that is

- 1/0.75*(ysb-yla): Transfer of carbon from the short-lived biota with dimensionless concentration ysb to the lower atmosphere with concentration yla with a time scale $\theta = 0.75$ yr corresponding to the approximate length of time of an agricultural crop or other short-lived (annual) plants.
- 1/150*(ylb-yla): Transfer of carbon from the long-lived biota with dimensionless concentration ylb to the lower atmosphere with a time scale $\theta = 150$ yr corresponding to the approximate length of time of forests of trees.
- 1/5*(yua-yla): Transfer of carbon from the upper atmosphere with dimensionless concentration yua to the lower atmosphere with a time scale $\theta = 5$ yr as estimated from earlier ODE models [2], [27] and other sources, e.g., [24].
- 1/30*(yul-yla): Transfer of carbon from the ocean upper layer with dimensionless concentration yul to the lower atmosphere with a time scale $\theta = 30$ yr as estimated from earlier ODE models [2], [27] and other sources, e.g., [24].

These values of θ_j^i are approximate estimates. Thus, a central question is how sensitive the model solutions (to Eqs. (3.3) to (3.9)) are to these values. To address this question, the reader can vary the values and observe the effect on the model solution, noting in particular the value of $y_{la}(t = 2013)$ which should be 400 as reported in Appendix D. An adjustment in c_1 of Eq. (3.10) programmed in model_1.R, model_1.m of Listings 3a, 3b may be required to meet this condition when parameters such as θ_j^i are changed.

APPENDIX C: CALCULATION OF pH

The fourth-order polynomial of Eq. (6.7) that is used to calculate $[H^+]$ (and thus pH) is derived starting with Eq. (6.6) for the total alkalinity, A

$$A = [HCO_3^-] + 2[CO_3^{2-}] + [B(OH)_4^-] + [OH^-] - [H]^+ \quad (6.6)$$

and the definitions of the equilibrium constants, Eqs. (6.1) to (6.5).

Solving for $[HCO_3^-]$ and $[CO_3^{2-}]$ in terms of p_m using Eqs. (6.2) and (6.3) gives

$$[HCO_3^-] = \frac{K_1[CO_2]}{[H^+]} = \frac{K_0 K_1 p_m}{[H^+]}, \quad (C1)$$

$$[CO_3^{2-}] = \frac{K_2[HCO_3^-]}{[H^+]} = \frac{K_0 K_1 K_2 p_m}{[H^+]^2}. \quad (C2)$$

Also, from Eq. (6.5) (with $B = [B(OH)_3 + [B(OH)_4^-])$)

$$[B(OH)_4^-] = \frac{B}{1 + [H^+]/K_B}. \quad (C3)$$

Substituting Eqs. (6.4), (C1), (C2) and (C3) in Eq. (6.6) gives

$$A = \frac{K_0 K_1 p_m}{[H^+]} + \frac{2K_0 K_1 K_2 p_m}{[H^+]^2} + \frac{BK_B}{K_B + [H^+]} + \frac{K_W}{[H^+]} - [H^+]$$

which can be rearranged as

$$K_0 K_1 p_m \left[H^+\right] \left(K_B + \left[H^+\right]\right)$$
$$+ 2K_0 K_1 K_2 p_m \left(K_B + \left[H^+\right]\right) + \left[H^+\right]^2 B K_B$$
$$+ K_W \left[H^+\right] \left(K_B + \left[H^+\right]\right) - \left[H^+\right]^3 \left(K_B + \left[H^+\right]\right)$$
$$- A \left[H^+\right]^2 \left(K_B + \left[H^+\right]\right) = 0.$$

Collecting coefficients for the various powers of $[H^+]$,

$$-1[H^+]^4 + (-K_B - A)[H^+]^3$$
$$+ (K_0 K_1 p_m + B K_B + K_W - A K_B)[H^+]^2$$
$$+ (K_0 K_1 p_m K_B + 2K_0 K_1 K_2 p_m + K_W K_B)[H^+]$$
$$+ (2K_0 K_1 K_2 p_m K_B) = 0.$$

Therefore, the coefficients of the fourth-order polynomial, Eq. (6.7), as listed in Table 5, are

$$[H^+]^0 : \quad D_0 = 2K_0 K_1 K_2 p_m K_B,$$
$$[H^+]^1 : \quad D_1 = K_0 K_1 p_m K_B + 2K_0 K_1 K_2 p_m + K_W K_B,$$
$$[H^+]^2 : \quad D_2 = K_0 K_1 p_m + B K_B + K_W - A K_B,$$
$$[H^+]^3 : \quad D_3 = -K_B - A,$$
$$[H^+]^4 : \quad D_4 = -1.$$

With $[H^+]$ available from factoring the fourth-order polynomial, pH can be calculated from Eq. (1.1).

APPENDIX D: *NEW YORK TIMES*, 10 MAY 2013 — ANNOUNCEMENT OF 400 ppm GLOBAL CO_2 CONCENTRATION

10 May 2013

Carbon Dioxide Level Passes Long-Feared Milestone
— Justin Gillis (NYT)

The level of the most important heat-trapping gas in the atmosphere, carbon dioxide, has passed a long-feared milestone, scientists reported on Friday, reaching a concentration not seen on the earth for millions of years.

Scientific monitors reported that the gas had reached an average daily level that surpassed 400 parts per million — just an odometer moment in one sense, but also a sobering reminder that decades of efforts to bring human-produced emissions under control are faltering.

The best available evidence suggests the amount of the gas in the air has not been this high for at least three million years, before humans evolved, and scientists believe the rise portends large changes in the climate and the level of the sea.

"It symbolizes that so far we have failed miserably in tackling this problem," said Pieter P. Tans, who runs the monitoring program at the National Oceanic and Atmospheric Administration that reported the new reading.

Ralph Keeling, who runs another monitoring program at the Scripps Institution of Oceanography in San Diego, said a continuing rise could be catastrophic. "It means we are quickly losing the possibility of keeping the climate below what people thought were possibly tolerable thresholds," he said.

The new measurement came from analyzers high atop Mauna Loa, the volcano on the big island of Hawaii that has long been ground zero for monitoring the worldwide carbon dioxide trend.

Devices there sample clean, crisp air that has blown thousands of miles across the Pacific Ocean, producing a record of rising carbon dioxide levels that has been closely tracked for half a century.

Carbon dioxide above 400 parts per million was first seen in the Arctic last year, and had also spiked above that level in hourly readings at Mauna Loa. But the average reading for an entire day surpassed that level at Mauna Loa for the first time in the 24 hours that ended at 8 p.m. Eastern Daylight Time on Thursday, according to data from both NOAA and Scripps.

Carbon dioxide rises and falls on a seasonal cycle and the level will dip below 400 this summer, as leaf growth in the Northern Hemisphere pulls about 10 billion tons of carbon out of the air. But experts say that will be a brief reprieve — the moment is approaching when no measurement of the ambient air anywhere on earth, in any season, will produce a reading below 400.

"It feels like the inevitable march toward disaster," said Maureen E. Raymo, a Columbia University earth scientist.

From studying air bubbles trapped in Antarctic ice, scientists know that going back 800,000 years, the carbon dioxide level oscillated in a tight band, from about 180 parts per million in the depths of ice ages, to about 280 during the warm periods between. The evidence shows that global temperatures and CO_2 levels are tightly linked.

For the entire period of human civilization, roughly 8,000 years, the carbon dioxide level was relatively stable near that upper bound. But the burning of fossil fuels has caused a 41% increase in the heat-trapping gas since the Industrial Revolution, a mere geological instant, and scientists say the climate is beginning to react, though they expect far larger changes in the future.

Governments have been trying since 1992 to rein in emissions, but far from slowing, emissions are rising at an accelerating pace, thanks partly to rapid economic growth in developing countries. Scientists fear the level of the gas could triple or even quadruple before being brought under control.

Indirect measurements suggest that the last time the carbon dioxide level was this high was at least three million years ago, during an epoch called the Pliocene. Geological research shows that the climate then was far warmer than today, the world's ice caps were smaller, and the sea level might have been as much as 60 or 80 feet higher.

Experts fear that humanity may be precipitating a return to such conditions — except this time, billions of people are in harm's way.

"It takes a long time to melt ice, but we're doing it," Dr. Keeling said. "It's scary."

Dr. Keeling's father, Charles David Keeling, began carbon dioxide measurements on Mauna Loa and at other locations in the late 1950s. The elder Dr. Keeling found a level in the air then of about 315 parts per million — meaning that if a person had filled a million quart jars with air, about 315 quart jars of carbon dioxide would have been mixed in.

His analysis revealed a relentless, long-term increase superimposed on the seasonal cycle, a trend that was dubbed the Keeling Curve. Subsequent research proved it was coming from the combustion of fossil fuels. Charles David Keeling died in 2005.

Countries have adopted an official target to limit the damage from global warming, which by most estimates requires that emissions stop by the time the level reaches about 450. "Unless things slow down, we'll probably get there in well under 25 years," Ralph Keeling said.

Yet many countries, including China and the United States, have refused to adopt binding national targets. Scientists say that unless far greater efforts are made soon, the goal of limiting the warming will become impossible without severe economic disruption.

"If you start turning the Titanic long before you hit the iceberg, you can go clear without even spilling a drink of a passenger on deck," said Richard B. Alley, a climate scientist at the Pennsylvania State University. "If you wait until you're really close, spilling a lot of drinks is the best you can hope for."

Climate-change contrarians, who have little scientific credibility but are politically influential in Washington, point out that carbon dioxide represents only a tiny fraction of the air — as of Thursday's reading, exactly 0.04%. "The CO_2 levels in the atmosphere are rather undramatic" a Republican congressman from California, Dana Rohrabacher, said in a Congressional hearing several years ago.

But climate scientists reject that argument, saying it is like claiming that a tiny bit of arsenic or cobra venom cannot have much effect. Research shows that even at such low levels, carbon dioxide is potent at trapping heat near the surface of the earth.

"If you're looking to stave off climate perturbations that I don't believe our culture is ready to adapt to, then significant reductions in CO_2 emissions have to occur right away," said Mark Pagani, a Yale geochemist who studies climates of the past. "I feel like the time to do something was yesterday."

This article has been revised to reflect the following correction:

Correction: May 10, 2013

An earlier version of this article misstated the amount of carbon dioxide in the air as of Thursday's reading from monitors. It is 0.04%, not 0.0004%.

APPENDIX E: EXERCISE PROBLEMS

A list of exercises follows. They are divided into two categories: (1) introductory exercises that illustrate basic aspects of the model of Eqs. (3.3) to (3.10) while requiring only small and readily programmed variations in the equations and/or associated routines, and (2) more advanced exercises that may require changes in the model equations and coding or additional mathematical analysis.

E1. Introductory Exercises

1. The sensitivity of the solution of Eqs. (3.3) to (3.10) to the model parameters, that is, $\theta_1, \ldots, \theta_7, c_1, r_1$, is of particular interest in finalizing the model and comparing the solution to any available observed data. To study this sensitivity, change the residence times for the lower atmosphere/upper ocean layer exchange, $1/5, 1/80$, (programmed as `dyla=1/5*(yua-yla)`, `dyul=1/80*(yla-yul)` in Listings 3a and 3b), by (a) 10%, (b) 100% and comment on the observed change in the numerical solutions from the base case (of Listings 1a, 1b, 5a and 5b). Note in particular that the observed value $y_{la}(t = 2013) = 400$ may not be maintained in the numerical output and to return it to this value (400) will require an adjustment in c_1 of Eq. (3.10) as programmed in Listings 3a and 3b.

2. Carry out a sensitivity analysis as outlined in the preceding exercise by varying one or more of the equilibrium constants programmed in the main program of Listings 1a and 1b.

```
k0=0.03347e-03;
k1=9.747e-04;
```

```
k2=8.501e-07;
kb=1.881e-06;
kw=6.46e-09;
```

Comment on the apparent sensitivity of the solution of Eqs. (3.3) to (3.9) to the values of these constants.

3. The numerical solutions discussed previously, e.g., in Figs. 2a and 2b, were computed with the ODE library integrators ode (R) and ode15s (Matlab) called in the main programs of Listings 1a and 1b.

```
#
# ODE integration
  out=ode(func=model_1,times=tout,y=y0);
```

```
%
% ODE integration
  reltol=1.0e-06;abstol=1.0e-06;
  options=odeset('RelTol',reltol,'AbsTol',abstol);
  [t,y]=ode15s(@model_1,tout,y0,options);
```

ode and ode15s are sophisticated ODE integrators that vary the integration step according to a user prescribed error tolerance. This is demonstrated by the call to ode15s above with specified relative and error tolerances of 10^{-6}. In the call to ode error tolerances are not specified and therefore ode uses the defaults which are also 10^{-6}.

Change the error tolerances for ode and ode15s and observe the effect on the numerical solutions. Comment: Since the solutions to Eqs. (3.3) to (3.9) are smooth exponentials (see Eq. (B6.2)), ode and ode15s should have little difficulty in achieving the specified accuracy and therefore changing the error tolerances is expected to have little effect on the numerical solutions.

The error tolerances for ode can be set as parameters, e.g., for the default tolerances

```
  out=ode(func=model_1,times=tout,y=y0,
          rtol=1.0e-06,atol=1.0e-06);
```

4. The calculation of $[H^+]$ by factoring the fourth order polynomial of Eq. (6.7) is based on the use of Newton's method of Eq. (6.8). Experience with this Newton iteration has indicated

that the calculation is very stable in the sense that few iterations are required to achieve convergence and there have been no convergence failures. To confirm this good numerical performance, add output statements to `spline.R`, `spline.m` of Listings 5a and 5b to display the details of the Newton iteration as indicated next.

R:

```
#
# End of while loop for Newton iteration
 }
```

Matlab:

```
%
% End of while loop for Newton iteration
  end
```

Add an output statement after the } or `end` to display:

(a) `iter`: Number of Newton iterations.

(b) `f`: Residual of the fourth order polynomial of Eq. (6.7).

(c) `df`: Derivative of the fourth order polynomial.

(d) `xhi`: Newton correction.

(e) `dif`: Change in $[H^+]$ from the last iteration.

Note in particular the small values of `iter` indicating rapid convergence of the Newton interations.

5. As the solution to Eqs. (3.3) to (3.9) proceeds through $1850 \leq t \leq 2100$ in the main programs of Listings 1a and 1b, compute Q_c at the output values of t ($t = 1850, 1860, \ldots, 2100$). Then at the end of the solution, plot $Q_c(t)$ against t to better understand how the source term Q_c in Eq. (3.4a) varies with t. This analysis of Q_c can also be done for `ncase=1,2,3,4` using the main programs in Listings 1c and 1d.

6. A portion of the CO_2 entering the lower atmosphere eventually is taken up by the oceans in three forms: (1) $CO_{2(aq)} + H_2CO_3$ (dissolved CO_2 plus carbonic acid), (2) HCO_3^- (bicarbonate ion) and (3) CO_3^{2-} (carbonate ion). The variation of these three species with t is explored graphically with `plot_out[8]=1` in the main programs of Listings 1a and 1b (option 8). Since the magnitudes of the three species differ significantly, they are plotted on a relative

basis using the $t = 1850$ values to normalize the variables on a common vertical (ordinate) scale.

However, the absolute values of the three species are also of interest. To investigate the absolute values, modify option 8 in plot_output.R of Listing 2a in the following way:

(a) The three ocean carbon species as a function of t.

```
#
# CO_2+H_2CO_3,HCO_3^{-},CO_3^{2-} (fractional change
# from 1850) vs ppm
  if(plot_out[8]==1){
  par(mfrow=c(1,1))
  plot(ppm,co2/co2[1],xlab="ppm",ylab="c(frac)",type="l",
    lwd=2,main="Fractional upper layer\n
    co2+h2co3,hco3,co3 vs cla (ppm)",ylim=c(0,3.5));
  points(ppm,co2/co2[1],pch="1",lwd=2);
   lines(ppm,hco3/hco3[1],type="l",lwd=2);
  points(ppm,hco3/hco3[1], pch="2",lwd=2);
   lines(ppm,co3/co3[1],type="l",lwd=2);
  points(ppm,co3/co3[1], pch="3",lwd=2);
#
# CO_2+H_2CO_3 vs t
  par(mfrow=c(1,1))
  plot(tout,co2,xlab="t",ylab="c(millimols/liter)",
    type="l",lwd=2,main="co2+h2co3 (millimols/liter)
    vs t");
#
# HCO_3^{-} vs t
  par(mfrow=c(1,1))
  plot(tout,hco3,xlab="t",ylab="c(millimols/liter)",
    type="l",lwd=2,main="hco3 (millimols/liter)
    vs t");
#
# CO_3^{2-} vs t
  par(mfrow=c(1,1))
  plot(tout,co3,xlab="t",ylab="c(millimols/liter)",
    type="l",lwd=2,main="co3 (millimols/liter)
    vs t");
}
```

Note that three plots have been added. The graphical output indicates (1) $CO_{2(aq)} + H_2CO_3$ increases with t as expected since atmospheric CO_2 is increasing with t, (2) HCO_3^- increases with t as might be expected, (3) CO_3^{2-} decreases somewhat unexpectedly with t. Thus, as the oceans become more acidic, the carbonate ion concentration decreases with important implications for coral and shelled animal life according to the reaction $CaCO_3(s) \rightleftharpoons Ca^{2+} + CO_3^{2-}$ ((s) designates solid). That is, as CO_3^{2-} decreases with increasing acidity, $CaCO_3(s)$ dissolves (the reaction shifts left to right).

Verify these effects by executing the preceding code for option 8 in `plot_output.R` of Listing 2a.

(b) To investigate how the three ocean carbon species vary with atmospheric CO_2, modify the three plots in the preceding exercise with `ppm` as the horizontal (abcissa) variable.

```
#
# CO_2+H_2CO_3 vs ppm
  par(mfrow=c(1,1))
  plot(ppm,co2,xlab="ppm",ylab="c(millimols/liter)",
    type="l",lwd=2,main="co2+h2co3 (millimols/liter)
    vs ppm");
#
# HCO_3^{-} vs ppm
  par(mfrow=c(1,1))
  plot(ppm,hco3,xlab="ppm",ylab="c(millimols/liter)",
    type="l",lwd=2,main="hco3 (millimols/liter)
    vs ppm");
#
# CO_3^{2-} vs ppm
  par(mfrow=c(1,1))
  plot(ppm,co3,xlab="ppm",ylab="c(millimols/liter)",
    type="l",lwd=2,main="co3 (millimols/liter)
    vs ppm");
```

(c) Modify the three plots in the two preceding exercises to investigate how the three ocean carbon species vary with ocean pH.

```
#
# CO_2+H_2CO_3 vs pH
```

```
par(mfrow=c(1,1))
plot(pH,co2,xlab="pH",ylab="c(millimols/liter)",
     type="l",lwd=2,main="co2+h2co3 (millimols/liter)
     vs pH");
#
# HCO_3^{-} vs pH
  par(mfrow=c(1,1))
  plot(pH,hco3,xlab="pH",ylab="c(millimols/liter)",
       type="l",lwd=2,main="hco3 (millimols/liter)
       vs pH");
#
# CO_3^{2-} vs pH
  par(mfrow=c(1,1))
  plot(pH,co3,xlab="pH",ylab="c(millimols/liter)",
       type="l",lwd=2,main="co3 (millimols/liter)
       vs pH");
```

Are the various plots (with t, ppm, pH as the horizontal variable) consistent with respect to variations in (1) $CO_{2(aq)} + H_2CO_3$, (2) HCO_3^- (3) CO_3^{2-}.

7. Verify that Eq. (B7.4) is the solution to Eq. (B7.2) (including the homogeneous IC $y(t = 0) = 0$).

8. Derive Eq. (B7.4) by: (1) separation of variables, (2) using an integrating factor, (3) the Laplace transform.

E2. More Advanced Exercises

1. Compute a numerical solution to Eq. (B7.2) and compare with the analytical solution Eq. (B7.4). Include the exact error (the difference between the numerical and analytical solutions) in the output.

2. The numerical integration of initial value ODEs is open ended with respect to t. To illustrate this, extend the basic time scale $1850 \le t \le 2100$ to $1850 \le t \le 2200$ in the main programs of Listings 1a and 1b and comment on the numerical solution, e.g., c_{la} in ppm and the ocean pH. Comment: Extension of the model by a century (or even centuries) is an extrapolation with increased uncertainty. A final value of $t = 2100$ in the base case of Listings 1a and 1b is a modest future projection.

3. Consider the coefficient matrix **c** of Eq. (B5.5). Compute and display the seven eigenvalues and associated eigenvectors of **c** using utilities available in R and/or Matlab.

   (a) Do the eigenvalues have the expected form for stable, non-oscillatory solutions. That is, are they real and nonpositive.
   (b) Change the residence times for the lower atmosphere/upper ocean layer exchange, $1/5, 1/80$, (programmed as `dyla=1/5*(yua-yla)`, `dyul=1/80*(yla-yul)` in Listings 3a and 3b), by (a) 10%, (b) 100% and comment on the observed change in the numerical eigenvalues and eigenvectors from the base case (of Listings 1a, 1b, 5a and 5b).

4. The equilibrium constants defined numerically in the main programs of Listings 1a and 1b are temperature dependent, and this effect may become increasingly important as the Earth warms. The equilibrium constants K_0, K_1, K_2, K_b as a function of temperature are given in [2], p. 131. For example,

$$-log(K_0) = \frac{-2622.38}{T} + 15.5873 - 0.0178471T$$
$$+ 0.0117950 \, Cl - 277676 \times 10^{-5} \, Cl \, T,$$

where Cl is the chlorinity, taken as 19.24 mg/ml, and T is absolute temperature, taken as 292.75°K ($= 19.59°$C). K_0 is of particular interest since it relates the atmospheric CO_2 partial pressure p_m to the ocean $[CO_2] = [CO_{2(aq)}] + [H_2CO_3]$, that is dissolved CO_2 plus carbonic acid H_2CO_3

$$K_0 = \frac{[CO_2]}{p_m}.$$

   (a) If the oceans eventually warm by 2°C, how will the equilibrium constant K_0 change.
   (b) Execute the model in Listings 1a, 1b, 5a and 5b with this changed K_0 to observe the effect on the equilibrium ocean chemistry from a 2°C temperature change. Take $K_w = 6.46 \times 10^{-9}$.

5. Program in Matlab the inclusion of the Mauna Loa data of Fig. 7 as discussed in Section 7.2. The programming in R can be used as a template.

6. Changes in the components of the ocean carbon can be compared by using fractional variables. The three components or chemical species are

$$CO_2 = CO_{2(aq)} + H_2CO_3, \ HCO_3^-, \ CO_3^{-2}.$$

The concentrations of these components can be computed using equilibrium chemistry based on the following constants

$$K_0 = \frac{[CO_2]}{p_m}, \qquad (2.6.1a)$$

$$K_1 = \frac{[HCO_3^-][H^+]}{[CO_2]}, \qquad (2.6.1b)$$

$$K_2 = \frac{[CO_3^{2-}][H^+]}{[HCO_3^-]}. \qquad (2.6.1c)$$

The sum of the three components is then (based on the notation of [2])

$$[CO_2] + [HCO_3^-] + [CO_3^{-2}]$$
$$= K_0 p_m + \frac{K_1 K_0}{[H^+]} p_m + \frac{K_2 K_1 K_0}{[H^+]^2} p_m = \frac{p_m}{\phi},$$

where

$$\frac{1}{\phi} = K_0 + \frac{K_1 K_0}{[H^+]} + \frac{K_2 K_1 K_0}{[H^+]^2}$$

or

$$\frac{1}{\phi} = K_0 \left(1 + \frac{K_1}{[H^+]} + \frac{K_2 K_1}{[H^+]^2} \right). \qquad (2.6.2)$$

Using the above relations, the fractional concentrations of the three species, f_{CO_2}, $f_{HCO_3^-}$ and $f_{CO_3^{2-}}$, can be expressed as follows.

$$f_{CO_2} = \frac{[CO_2]}{[CO_2] + [HCO_3^-] + [CO_3^{2-}]} = \phi K_0, \qquad (2.6.3a)$$

$$f_{HCO_3^-} = \frac{[HCO_3^-]}{[CO_2] + [HCO_3^-] + [CO_3^{2-}]} = \frac{K_1}{[H^+]} f_{CO_2} \qquad (2.6.3b)$$

$$f_{CO_3^{2-}} = 1 - f_{CO_2} - f_{HCO_3^-}. \qquad (2.6.3c)$$

Execute the following R code to compute the three fractional concentrations of Eqs. (2.6.3) for $4 \leq pH \leq 11$. Comment on how these fractions vary with increasing ocean acidification (decreasing pH). Note in particular that CO_3^{2-} decreases with increasing acidification. Thus for the reaction

$$CaCO_3(s) \rightleftharpoons Ca^{2+} + CO_3^{2-} \qquad (2.6.4)$$

as $[CO_3^{2-}]$ decreases with acidification, the reaction shifts left to right corresponding to a decrease in $CaCO_3$, that is, dissolution of $CaCO_3$.

```
#
# Equilibrium constants (Bacastow and Keeling)
  k0=0.03347;k1=9.75e-07;k2=8.50e-10;
#
# Number of output points
  n=15;
#
# Allocate arrays
  pH=rep(0,n);hion=rep(0,n);co2=rep(0,n);
          hco3=rep(0,n);co3=rep(0,n);
#
# Calculation of co2, hco3^{-}, co3^{2-}
# Step through pH
  for(i in 1:n){
#
#   H^{+} from pH (by definition of pH)
    pH[i]=4+0.5*(i-1);
    hion[i]=10^{-pH[i]};
    cat(sprintf("\n i = %2d  pH[i] = %4.1f
      hion[i] = %10.3e",
```

```
          i,pH[i],hion[i]));
#
#    Calculation of CO_2, HCO_3^{-}, CO_3^{2-}
     co2[i]=1/(1+k1/hion[i]+k1*k2/hion[i]^2);
     hco3[i]=k1*co2[i]/hion[i];
     co3[i]=1-co2[i]-hco3[i];
     cat(sprintf("\n          co2[i] = %10.3e
          hco3[i] = %10.3e    co3[i] = %10.3e\n",
          co2[i],hco3[i],co3[i]));
  }
#
# CO_2 vs pH
  par(mfrow=c(1,1))
  plot(pH,co2,xlab="pH",ylab="conc (normalized)",
     type="l",lwd=2,main="co2, hco3, co3 vs pH\n
     (normalized)");
  points(pH,co2, pch="1",lwd=2);
#
# HCO_3^{-} vs pH
  points(pH,hco3, pch="2",lwd=2);
   lines(pH,hco3,type="l",lwd=2);
#
# CO_3^{2-} vs pH
  points(pH, co3, pch="3",lwd=2);
   lines(pH, co3,type="l",lwd=2);
```

7. We consider a hypothetical scenario for some time in the future when the large industrialized states of Brazil, China, Europe, India, Russia and USA agree to collaborate on developing a practical fusion reactor. Eventually they succeed and, for the purpose of this exercise, the reactor is capable of providing pollution free energy. Subsequently, new fusion power stations start to be commissioned and come on stream. Their combined capacity increases steadily to produce energy sufficient to cause fossil fuel emissions to decrease at a constant linear rate.

  For this excercise, assume the fusion reactors start to come on line in 2040. Modify the ODE routine of Listing 3a so that the emissions rate of Eq. (3.10) falls linearly in t over a period of thirty years starting at 2040. Thus, the emissions rate Q_c will vary as

$$Q_c = c_1 \exp(-r_1(t - 1850)), \quad 1850 \le t \le 2040,$$

$$Q_c = c_1 \exp(-r_1(2040 - 1850))(1 - (t - 2040)/30),$$

$$2040 < t \le 2100.$$

(a) Plot atmospheric CO_2 concentration (y_{la} ppm) against time using `plot_out[1]=1` in the main program of Listing 1a.

(b) Plot the emission rate (Q_c) against time.

(c) What is the maximum level of CO_2 reached.

(d) What is $y_{la}(t = 2100)$ ppm.

8. Repeat the preceding exercise, but with fossil fuel emissions falling off exponentially with a time constant of $\tau = 20$ years. For this situation assume that from time 2040, the emissions rate will be attenuated by

$$\exp(-(t - 2040)/\tau), \quad 2040 < t \le 2100.$$

(a) Plot atmospheric CO_2 concentration (y_{la} ppm) against time using `plot_out[1]=1` in the main program of Listing 1a.

(b) Plot the emission rate (Q_c) against time.

(c) What is the maximum level of CO_2 reached.

(d) What is $y_{la}(t = 2100)$ ppm?

(e) Compare these results with those of the preceding exercise.

9. In order to ensure that the global mean temperature remains within acceptable bounds, it is necessary to ensure that CO_2 concentrations in the atmosphere do not exceed certain limits. Develop a mitigation strategy (e.g., exponential attenuation of the source term as $\exp(-(t - t_0)/\tau)$) that will ensure that atmospheric CO_2 concentration does not rise beyond a specified level. Do this by repeating the preceding exercise pertaining to fusion reactors with different starting times, t_0, for the fusion reactor (other than $t_0 = 2040$) so that the atmospheric CO_2 does not exceed $500, 600, 700, 800$ ppm.

10. Develop a mitigation strategy similar to the preceding exercise, but that also includes sequestration of CO_2 at deep levels in the oceans. For this exercise ignore all practical issues and assume that sequestration starts at year t_1 (to be specified) and ramps up linearly until it captures 20% of the emissions, $Q_c = c_1 \exp(-r_1(t - 1850))$ in Eq. (3.10), over a period of 10 years. This sequestration will require a positive source term in Eq. (3.8a) (for the ocean deep layer, dy_{dl}/dt), that is σQ_c where

$$\sigma = 0, \quad t < t_1,$$
$$\sigma = 0.2(t - t_1)/10, \quad t_1 \leq t \leq t_1 + 10,$$
$$\sigma = 0.2, \quad t > t_1 + 10.$$

Equation (3.4a) (for the atmosphere, dy_{la}/dt) will have a negative source term $-\sigma Q_c$. These source terms will be added to the programming of dy_{la}/dt and dy_{dl}/dt in Listing 3a.

11. Ocean acidification from CO_2 absorption has as a serious consequence the dissolution of coral and marine shell animals, with $CaCO_3$ as the principal component. The objective of this exercise is to calculate the maximum or equilibrium calcium ion concentration that can be held in solution for a given partial pressure, p_m, of $CO_{2(a)}$ above the solution ((a) denotes atmosphere), assuming the following species exist at equilibrium in solution:

$$Ca^{2+}, CO_3^{2-}, CO_2 \text{ (i.e., } CO_{2(aq)} + H_2CO_3), HCO_3^-, H^+.$$

The relation for the solubility of $CaCO_3$ is

$$CaCO_3(s) \rightleftharpoons Ca^{2+} + CO_3^{2-}. \tag{2.11.1}$$

Thus, the equilibrium Ca^{2+} concentration is the same as the $CaCO_3$ solubility in molar units.

The equilibrium solubility constant is given by

$$K_{sp} = [Ca^{2+}][CO_3^{2-}]. \tag{2.11.2a}$$

Equation (2.11.2a) comes from thermodynamics with the activity of solid $CaCO_3$ assumed to be unity and it equates the equilibrium between solid and dissolved calcium carbonate at saturation. In other words, at a given pH, calcium carbonate will either dissolve or precipitate to form the ionic species in Eq. (2.11.1) at concentrations that satisfy the solubility product condition of Eq. (2.11.2a).

As noted, the equilibrium Ca^{2+} concentration is the same as the $CaCO_3$ solubility. Thus knowing how the equilibrium concentration of the ionic species (in this case Ca^{2+}) changes with pH for a given partial pressure of $CO_{2(g)}$ above the solution, p_m, we can determine whether additional $CaCO_3$ would need to dissolve to maintain that equilibrium or, conversely, whether Ca^{2+} would need to precipitate (as $CaCO_3$) to maintain that equilibrium.

Thus, for the stated conditions, we need to determine the calcium ion concentration by combining the terms affected by the pH with the following relation (from Eq. (2.11.2a)),

$$[Ca^{2+}] = \frac{K_{sp}}{[CO_3^{2-}]} \qquad (2.11.2b)$$

From the equilibrium constant definitions, we have that

$$K_2 = \frac{[H^+][CO_3^{2-}]}{[HCO_3^-]} \qquad (2.11.3a)$$

or

$$[CO_3^{2-}] = \frac{K_2[HCO_3^-]}{[H^+]}. \qquad (2.11.3b)$$

Substituting Eq. (2.11.3b) in Eq. (2.11.2b)

$$[Ca^{2+}] = \frac{K_{sp}[H^+]}{K_2[HCO_3^-]}. \qquad (2.11.4)$$

Also

$$K_1 = \frac{[H^+][HCO_3^-]}{[CO_{2(aq)}]} \qquad (2.11.5a)$$

or

$$[HCO_3^-] = \frac{K_1[CO_{2(aq)}]}{[H^+]}. \qquad (2.11.5b)$$

Substituting Eq. (2.11.5b) in Eq. (2.11.4) gives

$$[Ca^{2+}] = \frac{K_{sp}[H^+]^2}{K_1 K_2 [CO_{2(aq)}]}. \qquad (2.11.6)$$

Finally,

$$K_0 = \frac{[CO_{2(aq)}]}{p_m} \qquad (2.11.7a)$$

or

$$[CO_{2(aq)}] = p_m K_0. \qquad (2.11.7b)$$

Substituting Eq. (2.11.7b) in Eq. (2.11.6) gives

$$[Ca^{2+}] = \frac{K_{sp}[H^+]^2}{K_0 K_1 K_2 p_m}. \qquad (2.11.8)$$

Equation (2.11.8) is now used to calculate $[Ca^{2+}]$ for a given $[H^+]$ and p_m. Because of the square dependence of the calcium ion concentration on hydrogen ion concentration, as the pH is decreased, i.e., the solution acidifies at a given p_m, the equilibrium calcium ion concentration rises rapidly, meaning more $CaCO_3$ will dissolve in order to maintain equilibrium. The reverse is true if the pH is increased, i.e., the solution is more basic for the given p_m. In this latter case, many fewer calcium ions can be dissolved and therefore $CaCO_3$ will precipitate. The actual p_m values will be determined from the solution of Eqs. (3.3) to (3.10) for a given set of initial conditions.

To illustrate the use of Eq. (2.11.8), from the model output in Table 3a at $t = 1850$ and the equilibrium constants of Listing 1a.

Variable	Value
pH	8.222
p_m	280.0
K_0	0.03347e − 03
K_1	9.747e − 04

$$K_2 \qquad 8.501\mathrm{e}-07$$

$$K_{sp}, S = 35 \qquad 4.27\mathrm{e}-07$$
$$(\mathrm{mol/lt})^2$$

$$4.27\mathrm{e}-04$$
$$(\mathrm{mol/lt})(\mathrm{mmol/lt})$$

A solubility constant (see Eq. (2.11.2a)), $K_{sp} = 4.27\mathrm{e}-04$ (mol/lt)(mmol/lt), is also included. The units reflect $[C_a^{2+}] \Rightarrow$ (mol/lt), $[CO_3^{2-}] \Rightarrow$ (mmol/lt).
The value for K_{sp} comes from the following calculation using R.

```
#
# Set salinity, temperature; calculate Ksp
  Ksp_S0 =KspC(S=0, TK=298.15)
  Ksp_S35=KspC(S=35,TK=298.15)
  cat(sprintf("\n Ksp_s0 = %9.3e, Ksp_s35 = %9.3e \n",
            Ksp_S0,Ksp_S35))
#
# Function KspC computes Ksp from the equation by Millero
  KspC=function(S,TK){
#    Millero (2006), pp 253-254
     pk=-171.9065-0.077993*TK+2839.319/TK+71.595*log10(TK)+
        (-0.77712+0.0028426*TK+178.34/TK)*sqrt(S)-
        0.07711*S+0.0041249*S^(3/2)
     Ksp=10^pk
     return(Ksp)}
```

The output from this code is

```
Ksp_s0 = 3.313e-09, Ksp_s35 = 4.272e-07
```

which demonstrates the strong dependence of K_{sp} on S (salinity). The equation for pK in the preceding code is from Millero [15], pp. 253–254.

We then insert these constants into Eq. (2.11.8) using R.

```
k0=0.03347e-03;k1=9.747e-04;k2=8.501e-07;ksp=4.27e-04;
hion=10^{-8.222+3};pm=280;
Ca=ksp*hion^2/(k0*k1*k2*pm);
cat(sprintf("\n Ca = %10.4e (mol/lt)\n",Ca));
```

Note the conversion from mol/lt to millimol/lt=mmol/lt in the pH calculation, a multiplication by 10^3,

```
hion=10^{-8.222+3}.
```

The output from this code is `Ca = 1.9782e-03 (mol/lt)`. Similarly, for $t = 2013$, from Table 3a, pH by linear interpolation is

$$pH = 8.100 + (8.086 - 8.100) \left(\frac{2013 - 2010}{2020 - 2010} \right) = 8.096,$$

```
k0=0.03347e-03;k1=9.747e-04;k2=8.501e-07;ksp=4.27e-04;
hion=10^{-8.096+3};pm=400.2;
Ca=ksp*hion^2/(k0*k1*k2*pm);
cat(sprintf("\n Ca = %10.4e (mol/lt)\n",Ca));
```

The output is `Ca = 2.4726e-03 (mol/lt)`. Thus, the increase in the Ca^{+2} saturation concentration is `1.9782e-03` (1850) to `2.4726e-03` (2013) due to a small decrease in pH, `8.222` to `8.096`.

As pH decreases with increasing t, e.g., the model base case in Table 3a, we would expect larger increases in the equilibrium $[Ca^{+2}]$. To investigate this point, perform the preceding calculations of $[Ca^{+2}]$ as a function of $[H^+]$ and p_m for the 26 values of t in Table 3a. In other words, add some R code to the main program of Listing 3a, including plotting of $[Ca^{+2}]$ vs pH.

Finally, we should keep in mind that the plot of $[Ca_2^+]$ vs pH is the same as plotting the $CaCO_3$ solubility vs pH. It might seem confusing at first since it looks like $CaCO_3$ is increasing strongly with decreasing pH but in fact the correct interpretation is the opposite, i.e., the $CaCO_3$ solubility is increasing with decreasing pH, and to the detriment of living things made of $CaCO_3$, i.e., they will dissolve as $CaCO_3$ solubility draws closer to the actual ocean $CaCO_3$. In other words, the decrease in the ocean $CaCO_3$ supersaturation[1] with decreasing pH is moving coral and shell animals closer to dissolution.

---

[1]Supersaturation applies to a solution containing an amount of a substance greater than that required for saturation. The saturation state Ω, of seawater

12. We know from carbonate chemistry, outlined in Section 6, that under equilibrium conditions there is a causal link between concentration of carbon dioxide in the atmosphere, p_m, and concentration of dissolved inorganic carbon, $DIC = T_c = CO_2 + HCO_3^- + CO_3^{-2}$, in the upper layer of the oceans. Here p_m represents the partial pressure of atmospheric carbon dioxide, which is very nearly equal to concentration in parts-per-million by volume, and $CO_2 = CO_{2(aq)} + H_2CO_3$. The evasion factor provides a simple way of determining the expected concentration change in $T_c = DIC$ due to changes in p_m.

For example, we can make the assumption that there is a power law relationship between partial pressure of carbon dioxide in the atmosphere, p_m, and concentration of dissolved inorganic carbon, DIC, (or total carbon, Tc). This can be expressed in the following mathematical form

$$DIC = cp_m^{1/\varepsilon},$$

where c is a constant and ε is the evasion factor discussed next. On taking logarithms we obtain

$$\log DIC = \log c + \frac{1}{\varepsilon}\log p_m. \qquad (2.12.1)$$

Then differentiating Eq. (2.12.1) with respect to p_m yields

$$\frac{1}{DIC}\frac{dDIC}{dp_m} = \frac{1}{\varepsilon}\frac{1}{p_m}. \qquad (2.12.2)$$

Rearranging and representing the derivative term by the ratio of small increments, we obtain a practical expression for calculating

---

with respect to $CaCO_3$, is determined from

$$\Omega = \frac{[Ca^{+2}][CO_3^{-2}]}{k_{sp}}.$$

A solution is in equilibrium when $\Omega = 1$ and is supersaturated when $\Omega > 1$. If Ω falls below 1 the solution is undersaturated and this can be caused by falling pH values. The effect of decreasing pH in the oceans is therefore to move coral and the shells of crusacean animals closer to dissolution.

the relative fractional change of p_m with respect to the fractional change of DIC, i.e.,

$$\varepsilon = \left(\frac{\delta p_m}{p_m}\right) \bigg/ \left(\frac{\delta DIC}{DIC}\right). \qquad (2.12.3)$$

The ratio ε is the so-called evasion factor, also known as the buffer factor. It is similar to a term used by Revelle and Suesse [23], later called the Revelle factor by Broecker and Peng [3]. Equation (2.12.3) is programmed in the main programs of Listings 1a and 1b, with increments δp_m and δDIC based on changes in 10 years (and thus a small increment for $1850 \leq t \leq 2100$).

Example 1

We first consider some basic carbon calculations for the atmosphere and oceans using selected output from the model (see Listings 1a and 1b). For example, we can calculate the mass of carbon in 1 m-a^3 (a pertains to air or atmosphere). Assuming that the atmosphere has a density of 1000 g-a/m-a^3 and molecular weight of 29 g-a/mol-a, the number of moles contained in 1 m-a^3 of air is equal to (1000 g-a/m-a^3)/(29 g-a/mol-a). Therefore, with carbon having a molecular weight of 12 g-c/mol-c, the atmosphere contained in year 1850

$$\frac{12 \text{ g-c/mol-c} \times 1000 \text{ g-a/m-a}^3}{29 \text{ g-a/mol-a}}$$

$$\times \frac{280}{10^0} \text{ mol-c/mol-a} = 0.1159 \text{ g-c/m a}^3.$$

Similarly, for 1 m-aq^3, in year 1850 the model gives $DIC = Tc = 1.7453$ mol-c/m-aq^3 (see Table 3a) for the ocean upper layer. Therefore, with the molecular weight of carbon being 12 g-c/mol-c, the ocean contains approximately

$$12 \text{ g-c/mol-c} \times 1.7453 \text{ mol-c/m-aq}^3 = 20.94 \text{ g-c/m-aq}^3.$$

Note that the two preceding results, 0.1159 g-c/m-a^3 and 20.94 g-c/m-aq^3, are based on different volumes, m-a^3 and m-aq^3, but the same carbon mass, g-c (rather than, for example, CO_2).

We can also use the equilibrium constants from Listings 1a and 1b for particular calculations. For example, using K_0,

$$CO_{2(aq)} = K_0 CO_{2(a)}$$
$$= (3.347 \times 10^{-5} \text{ millimol-c/lt/ppm})(280 \text{ ppm})$$
$$= (3.347 \times 10^{-5} \text{ mol-c/m-aq}^3/\text{ppm})(280 \text{ ppm})$$
$$= 0.0093716 \text{ mol-c/m-aq}^3$$

and the ratio of total carbon in the ocean, $CO_{2(aq)} + H_2CO_3 + HCO_3^- + CO_3^{2-}$, to $CO_2 = CO_{2(aq)} + H_2CO_3$ in the ocean is

$$\frac{1.7453 \text{ mol-c/m-aq}^3 (CO_{2(aq)} + H_2CO_3 + HCO_3^- + CO_3^{2-})}{0.0093716 \text{ mol-c/m-aq}^3 (CO_{2(aq)} + H_2CO_3)} = 185.67.$$

This means that a relatively small amount of CO_2 remains dissolved in the oceans. Rather, most of the carbon in the oceans takes the form of HCO_3^- and CO_3^{-2} through chemical reactions.

Example 2

Another important related concept is the uptake factor, due to Pilson [21], which is defined as the ratio of incremental change in DIC mol-c/kg-aq to the incremental change in p_m (atm), i.e.,

$$UF = \frac{\delta DIC}{\delta p_m} = \frac{1}{\varepsilon} \frac{DIC}{p_m} \text{ (mol-c/kg-aq)/atm-}CO_2. \quad (2.12.4)$$

Using the same data as above, i.e., $p_m = 280 \times 10^{-6}$ (atm-CO_2), $DIC = T_c = 1.7452$ mol-c/m-aq^3 and $\varepsilon = 8.95$ from Table 3a, Eq. (2.12.4) gives an uptake factor of

$$UF = \frac{1}{8.95} \frac{1.7453 \text{ (mol-c/m-aq}^3)/1000 \text{ kg-aq/m-aq}^3}{280 \times 10^{-6} \text{ atm-}CO_2}$$
$$= 0.696 \text{ (mol-c/kg-aq)/atm-}CO_2,$$

where we have converted DIC from (mol-c/m-aq^3) to (mol-c/kg-aq) based on the assumption that seawater has a density of approximately 1000 kg-aq/m-aq^3. Thus, if p_m increases from 280 ppm in year 1850 to 285.6 ppm in year 1860 (from Table 3a), i.e., $\delta p_m = 5.6$ ppm, we can expect that

$$\delta DIC = 0.696 \times 5.6 \times 10^{-6} \text{ atm}$$
$$= 3.93 \times 10^{-6} \text{mol-c/kg-aq} = 3.93 \times 10^{-3} \text{mol-c/m-aq}^3.$$

In other words, we would expect the concentration of DIC to increase by 0.00393 (mol-c/m-aq^3) from 1.7453 to 1.7492 (mol-c/m-aq^3), where this time we have converted back from mol-c/kg-aq to mol-c/m-aq^3. Thus, as expected, we have obtained the same value for DIC in 1860 as used above in Example 1 and appears in Table 3a.

Example 3

(a) Perform a hand calculation of $\varepsilon(t = 1860) = 8.880$ from Table 3a to verify the programming of ε in Listings 1a and 1b.

(b) Does the use of the arbitrarily selected value $\varepsilon(t = 1850) = 8.800$ appear consistent with the other values of ε in Table 3a for $t > 1850$.

(a) A plot of ε vs t is programmed in `plot_output.R` and `plot_output.m` of Listings 2a and 2b. For example, for the programming in R (of Listing 2a),

```
#
# eps vs t
  if(plot_out[2]==1){
  par(mfrow=c(1,1))
  plot(tout,eps,xlab="t",ylab="eps(t)",type="l",lwd=2,
      main="eps(t) vs t");
  }
```

Add a plot of ε vs *ppm* and a plot of ε vs *pH* for `plot_out[2]=1`. Note that as *ppm* increases and *pH* decreases (increased acidity), ε increases which reflects reduced ocean uptake of CO_2 with increasing atmospheric CO_2 and increasing acidity. Thus, there is a "bottleneck" for increased uptake of atmospheric CO_2 by the oceans.

13. Consider again the solubility product K_{sp} and the C_a^+ saturation Ω discussed in the preceding exercise. Add the following code to the end of the main program in Listing 1a in place of the call to plot_output.R of Listing 2a.

```
Main program of Listing 1a with call to plot_output
removed.
          .
          .
          .
#
# CO2 sensitivity (C^o/ppm CO2)
  co2s=0.005;
#
# Plot supersaturation
#
# ksp
  ksp1=4.27e-07           # [mol^2/kg^2] Millero, p254
  ksp=ksp1*10^6           # Converted to [mmol^2/lt^2]
#
# Current measured value of Ca converted to [mmol/lt]
  Ca_meas=1.028e-02*1000 # Millero, p62
#
# Ca at equilibrium conditions [mmol/lt]
  Ca_eq=ksp*hion^2/(k0*k1*k2*ppm);
#
# Supersaturation state
  Omega=Ca_meas*co3/ksp
  par(mfrow=c(3,1))
  par(mar=c(5,5,4,2)+0.1)
#
# Plot Omega vs ppm
  plot(ppm,Omega,type="l",lwd=3,col="blue",
    xlab="atmospheric, CO2 [ppm]",
    ylab=expression(paste(Omega,", [ - ]")),
    main="Ca supersaturation")
  grid(lty=1,col="darkgray")
#
# Plot Omega vs pH
  plot(pH,Omega,type="l",lwd=3,col="blue",
    xlim=rev(range(pH)),
    xlab="ocean upper layer, [pH]",
```

```
    ylab=expression(paste(Omega,", [ - ]")),
      main="Ca supersaturation")
   grid(lty=1,col="darkgray")
#
# Plot equilibrium Ca vs ppm
   plot(ppm,Ca_eq,type="l",lwd=3,col="blue",
      xlab="atmospheric, CO2 [ppm]",
      ylab="Ca, [mmol/lt]",
      main="Ca equilibrium")
   grid(lty=1,col="darkgray")
```

We can note the following details of the three plots produced by this code.

(a) Ω vs *ppm* indicates how the $CaCO_3$ supersaturation is decreasing as *ppm* increases (e.g., note the drop of $\Omega(t = 1850)$ to $\Omega(t = 2013)$).

(b) The drop in Ω with decreasing *pH* indicates the effect of increasing ocean acidification on increasing $CaCO_3$ dissolution (with serious consequences for coral and shelled marine life).

(c) The increase in saturation Ca^+ with *ppm* as Ω decreases.

Note that the vectorization facility of R has been used. For example, in

```
Ca_eq=ksp*hion^2/(k0*k1*k2*ppm);
```

hion and ppm are vectors with nout=26 elements as defined in the preceding code in the main program of Listing 1a.

As an exercise, vary (one at a time) the equilibrium constants K_0, K_1, K_2 and solubility product K_{sp} to observe how the three plots change with these constants. That is, do the plots pertaining to Ca^+ saturation depend markedly on the values of these constants.

Comment: Ω varies with location in the oceans. At some locations, $\Omega < 1$ in the deep ocean indicating a condition of undersaturation. Thus, Ω should be viewed as a worldwide average, with substantial variations.

14. Referring to the preceding exercise;

    (a) Translate the R code into Matlab and place it at the end of the main program of Listing 1b.

    (b) Execute the resulting Matlab (in Listings 1b, 2b, 3b, 4b and 5b) and compare the output with that of the R routines from the preceding exercise.

INDEX

A

acidity, *see* pH
acidification, 70, 124
alkalinity, 81–82, 84, 90, 181
analytical solution, *see* linear algebra
anthropogenic emissions, 5, 12, 15, 17, 60–66, 122
atmospheric CO_2, *see* CO_2

B

bicarbonate (HCO_3^-), 8, 17, 19, 22, 32–34, 43–44, 46–50, 71–74, 79–81, 89, 181, 189–191, 194–196
boron $(B(OH)_3, B(OH)_4^-)$, 17, 19, 21–22, 24, 27, 32–34, 43–44, 46–50, 79–80, 181

C

carbonate (CO_3^{2-}), 8, 17, 19, 22, 24, 32–34, 43–44, 46–50, 71–74, 79–81, 89, 181, 189–191, 194–196
carbonic acid (H_2CO_3), 8, 19, 22, 24, 32–34, 43–44, 46–50, 71–74, 79–81, 89, 181, 189–191, 194–196
calcium carbonate $(CaCO_3)$, 4, 195
  calcium ion concentration, 198–202
  chemistry, 4, 195
  coral, 202
  dissolution, 195, 198–202, 207–208
  shell animals, 202
  solubility product, 80, 198–202, 207–208
  supersaturation, 202, 207–208
calculus, fundamental theorem, x

carbon balances, x
characteristic equation, *see* linear algebra
chemistry, *see* ocean chemistry
clorinity, 80
climate change, ix
CO_2
  Carbon Dioxide Information Analysis Center, 2, 131
  comparison of model with observations, 125–130
  decreased ocean uptake, 71, 74
  dissolved, 71, 74, 80, 193, 203–206
  dynamics, 3
  effect of increasing atmospheric concentrations, ix, 183–185
  emission rate, 6, 12, 15, 17, 21, 32–33, 35, 99, 123
  function, 60–66, 99
  plotting, 170
  scenarios, 112–122, 196–197
  flux, 10
  increases, 183–186
  long term records, 184
  maximum level, 197
  measurement, 2, 125–130, 184
  mitigation, 198
  molecular weight, 30
  parts per million (ppm), *see* parts per million
  prediction, 2
  projection, 2, 183–186, 192
  sources of information, 2, 131
  water solubility, 80
computer routines, 15

APPENDIX: THE COMPANION MEDIA PACK

The Companion Media Pack for this book is available from the website of World Scientific Publishing. The download process is described in Section A.1 below. The media pack contains the Matlab and R source files for the models used in this book.

In the event that adjustments need to be made to the Media Pack's contents, we intend to provide that information and, if necessary, access to updated versions of the projects. Information on updating may be found on the World Scientific website or may be obtained by Email to the third author at wes1@lehigh.edu

A.1 Downloading the Media Pack

Begin the download process of obtaining the Companion Media Pack by visiting http://www.worldscientific.com/worldscibooks/10.1142/9516 and clicking on the 'Supplementary' tab. When you enter the site, you will be asked to login. If you have not previously registered on the World Scientific site, a simple (free) registration process will be required. Once you have successfully logged in and have reached the page for *An Introductory Global CO$_2$ Model*, select the 'Supplementary' tab. This tab contains a direct link to the Companion Media Pack. Clicking on the link will allow you to download the zipped file containing the source files provided by the authors. If, upon successful registration and login, you are redirected to another page, you may use the URL shown above to return to the correct page.

If your computer automatically places downloaded files in a folder named "Downloads", you will want to relocate the zipped file to a convenient permanent directory before unzipping.

Printed in the United States
By Bookmasters